BTEC Level 2

edexcel
advancing learning, changing lives

ENGINEERING LEVEL 2

BTEC First

Simon Clarke | Alan Darbyshire
Bill Mantovani | Bryan Weatherill
Series editor: Andy Boyce

A PEARSON COMPANY

Published by Pearson Education Limited, a company incorporated in England and Wales, having its registered office at Edinburgh Gate, Harlow, Essex, CM20 2JE. Registered company number: 872828

www.pearsonschoolsandfecolleges.co.uk

Edexcel is a registered trademark of Edexcel Limited

Text © Pearson Education Limited 2010
First published 2010

13 12 11 10
10 9 8 7 6 5 4 3 2 1

British Library Cataloguing in Publication Data
A catalogue record for this book is available from the British Library.

ISBN 978 1 84690 7234

Designed by Wooden Ark
Produced by Pearson
Typeset by HL Studios
Original illustrations © Pearson Education Limited/ HL Studios
Cover design by Visual Philosophy, Created by EMC Design
Index by Indexing Specialists (UK) Ltd
Cover photo © **Corbis**: Tim Pannell
Back cover photos © **Fotosearch**: tl; *Back:* **Alamy Images**: Peter Bowater tr; **Pearson Education Ltd**: Sophie Bluy br
Printed and bound in Great Britain at Scotprint, Haddington

Hotlinks
There are links to relevant websites in this book. In order to ensure that the links are up to date, that the links work, and that the sites are not inadvertently linked to sites that could be considered offensive, we have made the links available on the following website: www.pearsonhotlinks.co.uk. When you access the site, search for either the express code 7234V, title BTEC Level 2 First Engineering or ISBN 9781846907234.

Disclaimer
This material has been published on behalf of Edexcel and offers high-quality support for the delivery of Edexcel qualifications.
This does not mean that the material is essential to achieve any Edexcel qualification, nor does it mean that it is the only suitable material available to support any Edexcel qualification. Edexcel material will not be used verbatim in setting any Edexcel examination or assessment. Any resource lists produced by Edexcel shall include this and other appropriate resources.

The suggested activities in this book which involve practical work and exposure to hazards must be risk assessed and supervised by a qualified staff member.

Standards are current at the time of publication but may be subject to change during the life of the book. Please check with the relevant standards organisations.

Copies of official specifications for all Edexcel qualifications may be found on the Edexcel website: www.edexcel.com
Every effort has been made to contact copyright holders of material reproduced in this book. Any omissions will be rectified in subsequent printings if notice is given to the publishers.

Contents

Credits

The publisher would like to thank the following individuals for their help and advice on this project:

Brian Crossland and Mike Deacon

Picture credits

The publisher would like to thank the following for their kind permission to reproduce their photographs:

(Key: b-bottom; c-centre; l-left; r-right; t-top)

Alamy Images: Peter Bowater 129, Caro 181, chrisstockphoto 53, Alan Douglas 20b, Robert Fried 213, Tim Graham 87, ICP 42, 149, ICP 42, 149, imagebroker 207, Drive Images 135, Lenscap 245, Leslie Garland Picture Library 259, Horizon International Images Limited 144, milos luzanin 51, MBI 131, Stockbroker / MBI 210/2, Oramstock 139, Andrew Paterson 239, Chris Pearsall 29, Chris Howes / Wild Places Photography 257, qaphotos.com 1, Chris Rout 49, Chris Cooper-Smith 8, Image Source 11, 27, 85, 237, Image Source 11, 27, 85, 237, Image Source 11, 27, 85, 237, Image Source 11, 27, 85, 237, STOCKFOLIO® 263, StockImages 5, David J. Green - tools 187, Art Directors & TRIP 3, Brian Young 20t; **Andy Boyce:** 248l, 248c, 248r, 248b, 252, 257l, 257r, 262; **Fotosearch:** 202; **iStockphoto:** Elnur Amikishiyev 174c, **Zlatko Kostic** 179/2, Krzysztof Krzyscin 174l, **MACIEJ NOSKOWSKI** 138, Zhu ping 211, 224, Chris Schmidt 183, Jorgen Udvang 196, **Frank Wrigh**t 234; **Pearson Education Ltd:** Gareth Boden 89, 179, MindStudio 264, Jules Selmes 127, 155; **Photolibrary.com:** Fotosearch Value 153; **Shutterstock:** Monkey Business Images 151, Jakub Vacek 174r, Pedro Vidal 241

Cover images: *Front:* **Corbis:** Tim Pannell; **Fotosearch:** tl; *Back:* **Alamy Images:** Peter Bowater tr; **Pearson Education Ltd:** Sophie Bluy br

All other images © Pearson Education

Every effort has been made to trace the copyright holders and we apologise in advance for any unintentional omissions. We would be pleased to insert the appropriate acknowledgement in any subsequent edition of this publication.

About the Authors

Andy Boyce has 30 years of further education experience, teaching, assessing and managing engineering programmes from levels 2-5. He has recently worked as a unit specification writer for Edexcel. He has previously produced resources for BTEC in a box as well as materials to support levels 1, 2 and 3 Engineering and Manufacturing Diplomas (published by Pearson) and Edexcel down-loadable teacher support materials. For a considerable number of years he has worked as a BTEC External Verifier and is also a Principal Moderator for the Engineering Diploma.

He has provided input to the transition of the engineering programmes from NQF to QCC and is a BTEC Standards Verifier.

Simon Clarke has spent nearly 20 years delivering BTEC qualifications across First, National and Higher National Certificates and Diplomas. His lecturing experience in FE has led to his appointment as an Advanced Practitioner.
For the past 5 years he has been working with both schools and colleges as an external verifier for BTEC at level 2/3 as well as an External Examiner for Higher National Certificate/ Diploma and Degree programmes in Engineering subjects.
He has written extensively for textbooks, learning materials (including BTEC in a box) and unit specifications for Edexcel.

Alan Darbyshire

Alan Darbyshire works as an advisor for BTEC on their mechanical engineering programmes. Before his work for Edexcel he held the post of lecturer and senior lecturer in mechanical and plant engineering at the Blackpool and the Fylde College. He also spent 12 years working in the motor industry. He has written and co-written several engineering text books and tutor support materials.

Bill Mantovani has a long association with BTEC and delivering BTEC units but his involvement with electronics stretches back much further. He trained as an electronics engineer and he was a founder member of the team that developed the electronic and digital telephone system in use today. Bill joined the team delivering BTEC courses at Wakefield College in the 1980s. He was course tutor at a number of levels, including HNC/ HND and eventually took on all responsibilities for a number of HND courses run jointly with Sheffield Hallam and Leeds Metropolitan Universities.

Bill is currently involved with Edexcel First and National for Electrical and Electronic Engineering at Leeds City College and continues to write on his favourite subject of electronics and IT.

Bryan Weatherill began his career working in manufacturing engineering, before completing teacher training at Garnett College and teaching at North Devon College. As an Advanced Practitioner and Curriculum Leader he was involved with the introduction of BTEC Engineering courses. As Curriculum Leader, he selected the teaching team and carried out the training requirements for the BTEC ethos, a new style of teaching and learning for staff and students. He also created links with college feeder schools in their introduction of BTEC courses. He is a Registered Safety Practitioner and has specialised in the health and safety learning outcomes.
He currently works as an educational, engineering and health and safety consultant, in the UK and abroad.

About your BTEC Level 2 First in Engineering

Choosing to study for a BTEC Level 2 First Engineering qualification is a great decision to make for lots of reasons. This qualification will help you to build a foundation of knowledge in engineering, leading you into a whole range of professions or further study.

Your BTEC Level 2 First in Engineering is a **vocational** or **work-related** qualification. This doesn't mean that it will give you *all* the skills you need to do a job, but it does mean that you'll have the opportunity to gain specific knowledge, understanding and skills that are relevant to your future career.

What will you be doing?

The qualification is structured into **mandatory units** (ones you must do) and **optional units** (ones you can choose to do). This book covers 9 units in full – giving you a broad choice no matter what size your qualification.

- BTEC Level 2 First **Certificate** in Engineering: 1 mandatory unit **plus** optional units that provide a combined total of 15 credits

- BTEC Level 2 First **Extended Certificate** in Engineering: 2 mandatory units plus optional units that provide for a combined total of 30 credits

- BTEC Level 2 First **Diploma** in Engineering: 3 mandatory units, plus optional units that provide a combined total of 60 credits

The following table shows the units covered or partially covered in this book and whether they are Mandatory (M) or Optional (O) units for the Certificate, Extended Certificate and Diploma.

Unit number	Credit value	Unit name	Cert	Ex. Cert	Diploma
1	5	Working Safely and Effectively in Engineering	M	M	M
2	5	Interpreting and Using Engineering Information	O	M	M
3	5	Mathematics for Engineering Technicians	O	O	M
4	5	Applied Electrical and Mechanical Science for Engineering	O	O	O
8	5	Selecting Engineering Materials	O	O	O
10	10	Using Computer Aided Drawing Techniques in Engineering	O	O	O
14	10	Selecting and Using Secondary Machining Techniques to Remove Material	O	O	O
18	5	Engineering Marking Out	O	O	O
19	10	Electronic Circuit Construction	O	O	O

How to use this book

This book is designed to help you through your BTEC Level 2 First Engineering course. It is divided into 9 units.

This book contains many features that will help you use your skills and knowledge in work–related situations and assist you in getting the most from your course.

Introduction

These introductions give you a snapshot of what to expect from each unit – and what you should be aiming for by the time you finish it!

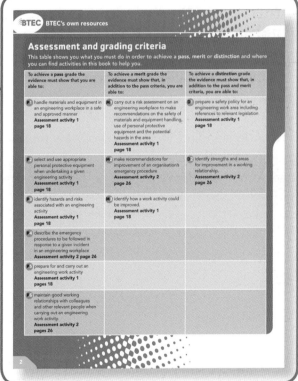

Assessment and grading criteria

This table explains what you must do in order to achieve the assessment criteria for each unit. For each assessment criterion in the table, shown by the grade button **P1**, there is an assessment activity.

Assessment

Your tutor will set **assignments** throughout your course for you to complete. These may take a variety of forms, from research, presentations, performances and evaluations to reports and posters.
The important thing is that you evidence your skills and knowledge to date.

Stuck for ideas? Daunted by your first assignment? These students have all been through it before…

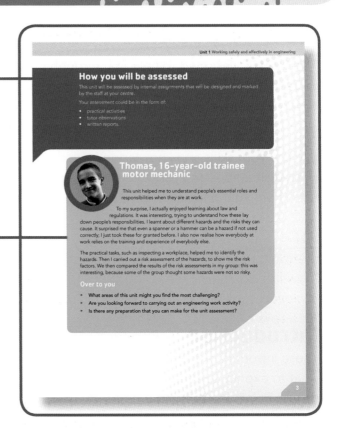

Activities

There are different types of activities for you to do: **assessment activities** are suggestions for tasks that you might do as part of your assignment and will help you demonstrate your knowledge, skills and understanding. Where appropriate these have **grading tips** that clearly explain what you need to do in order to achieve a merit or distinction grade.

There are also suggestions for activities that will give you a broader grasp of the industry, stretch your imagination and deepen your skills.

Activity: HASAWA

The table below shows some sections of HASAWA. Research the Act and indicate by ticks which sections apply to the persons listed.

	Employer	Employee
Section 2		
Section 3		
Section 7		
Section 8		

Personal, learning and thinking skills (PLTS)

Throughout your BTEC Level 2 Engineering course, there are lots of opportunities to develop your personal, learning and thinking skills. Look out for these as you progress.

PLTS

By exploring hazards and risks associated with an engineering activity you will develop your skills as an **independent enquirer**.

Functional skills

It's important that you have good English, maths and ICT skills – you never know when you'll need them, and employers will be looking for evidence that you've got these skills too.

Functional skills

By writing a report to describe the emergency procedures to be followed in response to an incident, you will develop your **writing** skills.

Key terms

Technical words and phrases are highlighted in bold, If the word appears in blue you will find the definition in a Key term box on the page. If the bold word or term appears in black you will be able to find the definition in the glossary.

Key term

PPE – personal protective equipment, worn as appropriate for the task being carried out. Examples: overalls, protective footwear (such as steel toe caps), eye protection, dust masks and respirators to cover the mouth and nose.

WorkSpace

Case studies provide snapshots of real workplace issues, and show how the skills and knowledge you develop during your course can help you in your career.

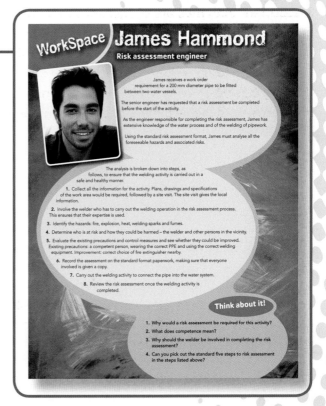

Just checking

When you see this sort of activity, take stock! These quick activities and questions are there to check your knowledge. You can use them to see how much progress you've made or as a revision tool.

Edexcel's assignment tips

At the end of each chapter, you'll find hints and tips to help you get the best mark you can, such as checklists to help you remember processes and really useful facts and figures.

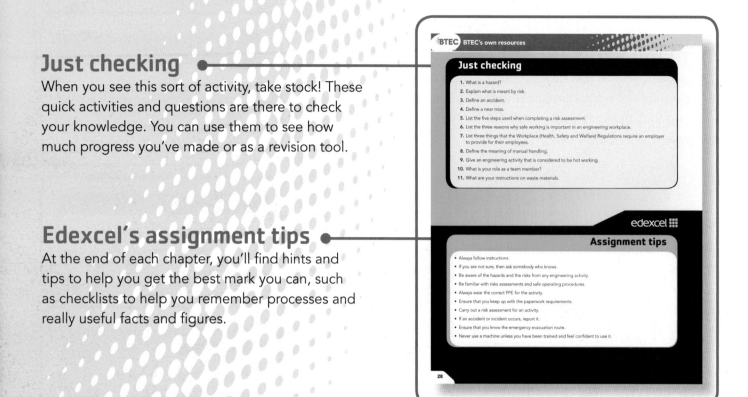

BTEC BTEC's own resources

Just checking

1. What is a hazard?
2. Explain what is meant by risk.
3. Define an accident.
4. Define a near miss.
5. List the five steps used when completing a risk assessment.
6. List the three reasons why safe working is important in an engineering workplace.
7. List three things that the Workplace (Health, Safety and Welfare) Regulations require an employer to provide for their employees.
8. Define the meaning of manual handling.
9. Give an engineering activity that is considered to be hot working.
10. What is your role as a team member?
11. What are your instructions on waste materials.

edexcel

Assignment tips

- Always follow instructions.
- If you are not sure, then ask somebody who knows.
- Be aware of the hazards and the risks from any engineering activity.
- Be familiar with risks assessments and safe operating procedures.
- Always wear the correct PPE for the activity.
- Ensure that you keep up with the paperwork requirements.
- Carry out a risk assessment for an activity.
- If an accident or incident occurs, report it.
- Ensure that you know the emergency evacuation route.
- Never use a machine unless you have been trained and feel confident to use it.

28

Have you read your BTEC Level 2 First Study Skills Guide? It's full of advice on study skills, putting your assignments together and making the most of being a BTEC Engineering learner.

Your book is just part of the exciting resources from Edexcel to help you succeed in your BTEC course. Visit www.edexcel.com/BTEC or www.pearsonschoolsandfecolleges.co.uk/BTEC for more details.

Working safely and effectively in engineering

1

Everyone who works in engineering must be able to perform safely and healthily. To ensure everyone's wellbeing, including your own, you must be able to spot potential hazards and risks.

This unit starts by exploring the importance of working within relevant legislation and adopting safe practices. We'll also look at typical incidents that an engineer has to deal with, such as contacting first-aiders following an accident and sounding alarms.

The unit provides you with the opportunity to develop the skills and understanding you need to perform a range of engineering tasks safely. It also looks at working relationships, focusing on the importance of cooperation. The ability to build good working relationships is part of the toolkit for everyone working in engineering.

You'll be able to transfer the skills and knowledge that you gain through studying this unit to other parts of your studies and to your everyday working life.

Learning outcomes

After completing this unit you should be able to:

1. apply statutory regulations and organisational safety requirements
2. work efficiently and effectively in engineering.

Assessment and grading criteria

This table shows you what you must do in order to achieve a **pass**, **merit** or **distinction** and where you can find activities in this book to help you.

To achieve a **pass** grade the evidence must show that you are able to:	To achieve a **merit** grade the evidence must show that, in addition to the pass criteria, you are able to:	To achieve a **distinction** grade the evidence must show that, in addition to the pass and merit criteria, you are able to:
P1 handle materials and equipment in an engineering workplace in a safe and approved manner **Assessment activity 1 page 18**	**M1** carry out a risk assessment on an engineering workplace to make recommendations on the safety of materials and equipment handling, use of personal protective equipment and the potential hazards in the area **Assessment activity 1 page 18**	**D1** prepare a safety policy for an engineering work area including references to relevant legislation **Assessment activity 1 page 18**
P2 select and use appropriate personal protective equipment when undertaking a given engineering activity **Assessment activity 1 page 18**	**M2** make recommendations for improvement of an organisation's emergency procedure **Assessment activity 2 page 26**	**D2** identify strengths and areas for improvement in a working relationship. **Assessment activity 2 page 26**
P3 identify hazards and risks associated with an engineering activity **Assessment activity 1 page 18**	**M3** identify how a work activity could be improved. **Assessment activity 1 page 18**	
P4 describe the emergency procedures to be followed in response to a given incident in an engineering workplace **Assessment activity 2 page 26**		
P5 prepare for and carry out an engineering work activity **Assessment activity 1 page 18**		
P6 maintain good working relationships with colleagues and other relevant people when carrying out an engineering work activity. **Assessment activity 2 page 26**		

How you will be assessed

This unit will be assessed by internal assignments that will be designed and marked by the staff at your centre.

Your assessment could be in the form of:

- practical activities
- tutor observations
- written reports.

Thomas, 16-year-old trainee motor mechanic

This unit helped me to understand people's essential roles and responsibilities when they are at work.

To my surprise, I actually enjoyed learning about law and regulations. It was interesting, trying to understand how these lay down people's responsibilities. I learnt about different hazards and the risks they can cause. It surprised me that even a spanner or a hammer can be a hazard if not used correctly; I just took these for granted before. I also now realise how everybody at work relies on the training and experience of everybody else.

The practical tasks, such as inspecting a workplace, helped me to identify the hazards. Then I carried out a risk assessment of the hazards, to show me the risk factors. We then compared the results of the risk assessments in my group: this was interesting, because some of the group thought some hazards were not so risky.

Over to you

- What areas of this unit might you find the most challenging?
- Are you looking forward to carrying out an engineering work activity?
- Is there any preparation that you can make for the unit assessment?

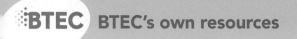

1.1 Be able to apply statutory regulations and organisational safety requirements

Start up

Why work in a safe and healthy manner?

Think of an engineering activity that you know is hazardous. For example:

- working with tools
- working with machinery
- working with hazardous materials
- manual handling.

List five hazards within your chosen engineering activity. For each **hazard**, consider:

- the possible effects of carrying out the activity
- the training that people need
- any psychological factors.

Discuss your findings in small groups. Compare the factors you have identified from your chosen activity with those of other engineering activities.

PLTS

By exploring hazards and risks associated with an engineering activity you will develop your skills as an **independent enquirer**.

Discussing your findings with other members of your group will help you develop your skills as a **team worker**.

1.1.1 Materials and equipment handling

The main factor in health and safety is to ensure that no person in an engineering environment is injured or suffers ill health.

There are three good reasons why safe working is important in an engineering workplace:

- The cost of accidents can be very high. It includes accident investigation, repairs to machinery, compensation payments to injured people, legal costs and increased insurance costs.
- Moral reasons: people's lives and wellbeing can be affected by what happens to them at work. Everybody has a moral obligation not to cause harm to others. It is not acceptable to put people at **risk** by having poor health and safety systems. Untrained workers are not aware of the dangers in carrying out work.

- Legal reasons: health and safety legislation places duties on everyone involved in engineering activities. These duties are the responsibilities of **employers**, of employees and of any other person affected.

Roles and responsibilities

Everybody within any engineering business has health and safety responsibilities. These are clearly outlined in:

- the Health and Safety at Work etc Act 1974
- the employer's health and safety policy
- other current and relevant legislation, such as the Management of Health and Safety at Work Regulations 1999, the Workplace (Health, Safety and Welfare) Regulations 1992 and the Personal Protective Equipment at Work Regulations 2002.

The employer's health and safety policy

A health and safety policy is a business plan for safety. You will find that all engineering organisations have such a policy, because it is a legal requirement.

A policy should include three things:

- Statement of intent. This is one A4 page that sets out the health and safety aims and objectives. The employer shows commitment to the health and safety policy by the Managing Director signing the statement of intent.

- Organisation. This describes how health and safety is organised, by showing a structure of responsibility for all employees. It determines the use of competent **employees** by allocating their responsibilities and accountability. It requires the cooperation of all employees through consultation. The communication lines are required to be shown through the structure.

- Arrangements. These are the 'nuts and bolts' of the policy. Included in this part are the procedures and systems for first aid, risk assessments, welfare requirements and consultation with employees. These and many others are legal requirements of the various regulations and the Health and Safety at Work etc Act (see below).

The Health and Safety at Work etc Act (HASAWA) 1974

The Health and Safety at Work Act is a United Kingdom Act of Parliament. This means that it has progressed through the parliamentary system and is now law. Anyone who does not comply with the Act is breaking the law. People who break that law can be taken before the courts and be punished by being fined, or even sent to prison.

Key terms

Hazard – Something with the potential to cause harm to you or to someone else. Example: handling a hazardous substance, such as an acid, without wearing the correct personal protective equipment.

Risk – The likelihood that harm will occur and the severity of the harm from working with the hazard.

Employer – the person/s that run, control and manage a workplace where engineering activities are carried out. Examples: a factory, a manufacturing plant, a construction site.

Employee – a person who is employed, through a contract of employment, to carry out specific jobs, tasks or activities for an employer. Examples: designers, motor vehicle technician fitters, electricians.

 Did you know?

To visit a website on Health and Safety visit the hotlinks section on page ii.

Copies of the Health and Safety at Work Act and Regulations can be obtained directly from the HSE, ordered from some book stores, or borrowed from local libraries.

Key terms

Did you know?

Most organisations use the figures 1, 2, 3, 4 and 5 to denote both the probability of an accident happening and the severity of the resulting injury. With 1 being the lowest in each case. The multiplying of the numbers during the risk assessment process gives a risk priority indicator. 1 × 1 is very low risk, but 5 × 5 is very high risk.

The responsibilities of an employer under the Act are to provide their employees with:

- a **safe system of work**
- a safe and healthy workplace
- safe **work equipment**
- safe methods of storing, transporting, handling, using and disposing of substances and materials
- **competent**, properly trained supervisors.

Under the Act, your responsibilities as an employee are:

- to cooperate with the employer in all matters regarding health and safety
- not to put yourself or other people at risk
- not to misuse or interfere with anything relating to health and safety
- to report defects or dangerous situations that you find in the workplace
- to work safely, following instructions and training.

Activity: HASAWA

The table below shows some sections of HASAWA. Research the Act and indicate by ticks which sections apply to the persons listed.

	Employer	Employee
Section 2		
Section 3		
Section 7		
Section 8		

Regulations

Regulations are derived from European Directives. Because the United Kingdom has been a member of the European Union since 1963, these Directives have to be implemented in UK law. This is to ensure 'harmonisation and fair trading' across all EU member states. Regulations are secondary to the Health and Safety at Work Act, so anyone who does not comply with them is breaking the law. They can then be prosecuted through UK criminal law.

The Management of Health and Safety at Work Regulations (MHSW) 1999

These regulations apply to all workplaces, including those carrying out engineering activities. The regulations recognise that the employer has the main responsibility for health and safety and that the employee also has responsibilities.

Regulation 3 requires that employers carry out 'suitable and sufficient' risk assessments for all activities. To lessen the risks, they have to put precautions in place so that you, as an employee, are not injured.

The aim of risk assessments is to identify hazards so that action can be taken to eliminate, reduce or control them and accidents can be prevented.

Regulation 4 requires that employers integrate health and safety into the management systems of their companies. This is to ensure that you as an employee are protected by precautionary measures that are properly planned and organised.

The Workplace (Health, Safety and Welfare) Regulations (WHSWR) 1992

The Workplace (Health, Safety and Welfare) Regulations 1992 are concerned with general safety in an engineering workplace. They are to ensure that as an employee you can enter and exit from any part of the workplace in a safe manner. The Regulations also require that:

- all **fixtures and fittings** are well maintained and kept clean
- the workplace atmosphere remains healthy through an adequate supply of fresh air
- temperatures are maintained at a minimum of 16°C for normal work and 13°C for energetic work
- lighting is of the correct level for work being carried out, including lighting required for emergency evacuations
- all work areas are kept clean; this is known as housekeeping
- measures are taken to prevent falls from height where they are likely to cause injury to employees
- traffic management in the work area ensures that vehicles and pedestrians are segregated
- welfare facilities are provided so that you as an employee can use toilets and wash and dry their hands
- fresh drinking water is provided when required.

Key terms

Fixtures and fittings – such items as electrical wiring, lighting, doors, ventilation systems, windows, carpets.

Emergency lighting – the green and white signs that indicate the evacuation route in an emergency.

This welder is wearing gloves to protect his hands and a face mask to protect his face and eyes – but what about his arms?

Key term

PPE – personal protective equipment, worn as appropriate for the task being carried out. Examples: overalls, protective footwear (such as steel toe caps), eye protection, dust masks and respirators to cover the mouth and nose.

PLTS

By exploring hazards and risks associated with an engineering activity you will develop your skills as an **independent enquirer**.

Personal Protective Equipment at Work Regulations 1992 (as amended 2002)

The employer must provide personal protective equipment to protect individual workers. For example, if you are asked to use a grinding machine to sharpen a chisel, you must wear safety glasses or goggles to avoid injuries to your eyes.

The choice of personal protective equipment is based on:

- the type of hazard from the activity being carried out
- the correct material for the **PPE**. For example, a welder would wear a flameproof boiler suit.

PPE must fit the person wearing it and be comfortable to wear. For example, the flameproof boiler suit must not be too small or too large and must be made from 'breathable' material.

PPE must be supplied free of charge to you as an employee and must be cleaned, repaired and replaced when not fit for use.

You as an employee have the right under the Regulations to have a say in the choice of the PPE and to be trained in its correct use.

Activity: Selection of PPE

List the items of personal protective equipment that a tyre fitter, a lathe operator or a motor vehicle technician would wear when working.

Manual Handling Operations Regulations (MHOR) 1992

One third of all reported injuries are due to incorrect **manual handling** methods. Most engineering activities involve some form or other of manual handling. When working in engineering you will be instructed and trained in the correct methods of manual handling. The instruction and training will usually be carried out in the first week as part of your **induction** programme into a company.

Employers should ensure that their employees avoid manual handling of items that could injure them. If manual handling cannot be avoided, then a manual handling assessment must be carried out. This is in four parts:

1. The task or activity being carried out. For example, imagine that you have to carry out maintenance on a machine. This would involve you lifting and moving parts of machinery.

   Assess the manual handling job that you have been asked to do.

Does it need to be done?

Can it be made safer, by splitting into smaller parts?

2. The capacity of the individual doing the work.

    Are you fit and healthy?

    Have you any existing back, arm or shoulder problems?

    Have you been trained in manual handling?

3. The load.

    Are parts of the machine greasy, dirty or hot? Do they have sharp edges?

    Can you make the load smaller, lighter, or more secure?

    Can you make it easier by ensuring a better grip, or by using some handles?

    Can you put the load into a container so that it is more stable?

4. The working environment.

    Is there sufficient space to work in?

    Is there enough light?

    Look around: are there any obstructions on the route where you are to carry the load?

    Are there any changes in floor level, such as steps or stairs?

    Are there any spills of water, oils or other substances on the floor?

Key terms

Manual handling – the transporting or supporting of a load (including lifting, putting down, pushing, pulling, carrying, or moving) by hand or by bodily force.

Induction – the first few days of starting at a new place of work.

Activity: Reading drawings

Work in pairs or small groups.

1. Use the Internet or library to research the kinetic lifting technique.
2. Together prepare a poster to explain this, showing diagrams.

PLTS

Discussing your findings with other members of your group will help you develop your skills as a **team worker**.

Control of Substances Hazardous to Health Regulations 2002 (COSHH)

COSHH Regulations must be complied with, to ensure the safe use and handling of hazardous chemicals in engineering workplaces. This will reduce the likelihood of your suffering any ill-health effects from working with the chemicals.

Some chemicals, such as petrol and oils, can be absorbed into your body through your skin and eyes. Others can enter your body through cuts and grazes or through accidental injection from chemically contaminated pieces of sharp metal.

Chemicals can also be swallowed. This could happen if you eat or drink while handling or using hazardous chemicals.

Gases, dusts and fumes are airborne chemicals that you could breathe into your lungs. Asbestos, when breathed in, can stay in the lungs and cause disease.

Health and Safety (Safety Signs and Signals) Regulations 1996

When you work in engineering you must be familiar with the five categories of standard safety signs:

- **mandatory** – signs that show that we have to do something. These show a requirement for human behaviour, for example, wearing ear defenders in a noisy environment.

- **warning** – signs that give us a warning of a hazard or danger. We must take care and follow the precautions shown. A black flash shows an electrical hazard.

- **prohibition** – these signs are intended to stop us behaving in a dangerous manner. For example, do not enter a hazardous area.

- **fire** – these signs show us the location and type of fire fighting equipment.

- **safe condition** – signs that give us information on emergency exits, first aid facilities and rescue equipment.

The Chemical (Hazard Information and Packaging) Regulations (CHIP) 2009

All chemicals and substances for use in any engineering workplace must be labelled and safety information must be supplied to comply with these regulations. It is important that you, as an employee working in that environment, are aware of the hazards of each chemical and substance that you come into contact with.

Where a hazardous chemical is supplied in any sort of package, the package must be labelled. The packaging must be strong and made

---

Remember

When you are using or working with chemicals, note that:

- some chemicals are more hazardous than others

- some will harm you faster than others

- you should always wash your hands before eating or drinking

- you should always use the correct type of PPE.

of the correct type of material for the chemical. It must be able to withstand the conditions of transporting, use and storage. The label must state the hazards from the chemical and the precautions to be taken when handling it.

Regulation 2 of CHIP defines the categories of hazards for chemicals and substances as one of the following:

Physio-chemical properties:

- **explosive** (e.g. ammonium nitrate used as a fertiliser)

- **oxidising** (e.g. hydrogen peroxide used by hairdressers as a bleaching agent for hair)

- highly **flammable** (e.g. liquefied petroleum gas; butane or propane used for heating and cooking)

- flammable (e.g. diesel fuel for car and truck engines).

Health effects:

- very **toxic** (e.g. potassium cyanide, a chemical used during electroplating)

- toxic (e.g. lead paint, found on woodwork in older properties)

- harmful if swallowed or on contact with skin (e.g. petrol or anti freeze found in vehicle engines)

- **corrosive** (e.g. acid used as an electrolyte in vehicle batteries)

- **irritant** (e.g. bleach, soaps, washing powders used in cleaning)

- **sensitising** (e.g. cement dust, hard wood dusts found in the construction industry)

- **carcinogenic** (e.g. used oils from vehicle engines).

1.1.2 Using equipment safely

Mechanical safety

Rotating shafts, belts, gears and work-holding equipment are hazardous parts of machinery. In general, in engineering workshops all these hazardous parts should be guarded.

If you are asked to operate any machine, you must be confident in its operation. This is important. You as the operator must have received training on the machine and in what it does. Part of this training would be to ensure that you understand the hazards from the machine.

Electrical safety

Electricity provides the energy used to power machinery, lighting and work equipment in an engineering environment. Electricity flows through circuits and is easily controlled by switches and other electrical apparatus to ensure that the machinery and lighting work efficiently and safely.

Key terms

Explosive – able or likely to explode.

Oxidising – the combination of a substance with oxygen. Substances such as these may increase the risk and violence of a fire significantly if they come into contact with a flammable or combustible substance.

Flammable – a substance that burns easily.

Toxic – a substance containing poison or caused by poisonous substances.

Corrosive – a substance that destroys living tissues.

Irritant – a substance that causes inflammation on contact with skin.

Sensitising – adverse effects caused by a reaction to inhalation or penetration of a substance.

Carcinogenic – a substance that can cause cancer if inhaled, ingested or if penetrates the skin.

This machine tool operator is wearing safety goggles and the machine has a guard to protect him from swarf flying off

Any contact with an electrical conductor will result in an **electrical shock**, which may harm you.

Remember; it is essential to isolate any electrical equipment before working on it.

Fluid power equipment

Fluid power can be provided by compressed air or hydraulic oil.

Compressed air is often used in assembly equipment: the air pressure drives pistons in cylinders to give either straight-line movements or rotary movements.

For example, a dentist will clean your teeth using an air-driven rotary tool with different attachments fitted to the tool end. Compressed air is also used in the food industry to mix foods and then package them.

Engineers are required to install, maintain and repair machinery such as this. Injuries can be caused if you come into contact with a release of compressed air, or are struck by one of the moving pistons.

Hydraulic systems are used for more heavy-duty work. Earth-moving equipment and hoists are powered by hydraulic systems. The fluid is pumped through a series of valves that control flow into the cylinders to move the linear pistons or rotary motors.

The hydraulic fluid is not compressible and so it exerts tremendous pressures into the moving parts of machines. Anybody coming into contact with an escape or leak or this high-pressure fluid can suffer serious injuries. There is also the risk of serious crush injuries from the piston movements and the fluid itself can cause cancers on the skin.

Housekeeping

This is the term used to describe the way a person works and the general state of the surrounding work area. It covers:

- keeping tools clean and stored in the correct place when not in use
- ensuring that all waste materials are put into the correct containers
- keeping the work area clear so that you can work safely
- making sure that all access areas, gangways and fire doors are kept clear
- ensuring that any spillages of water, oils or chemicals are cleaned up straight away
- ensuring that you protect yourself and other persons from injury and ill-health when carrying out engineering activities.

1.1.3 Hazards and risks

The risk assessment process

When you are working in engineering, you are required to avoid harming or causing ill-health both to yourself and to other people. You need to be aware of the hazards that can be found in an engineering workplace. To achieve this you need to carry out a **risk assessment**.

There are five steps in a risk assessment, as follows.

1. Identify the hazards from the engineering activity. Hazards fall into the following general types:

    - mechanical, e.g. working machinery

    - physical, e.g. wet floors

    - chemical, e.g. contact with a corrosive chemical

    - biological, e.g. contact with another person's body fluids

    - ergonomic, e.g. repetitive or strenuous movements of parts of the body

    - psychological, e.g. pressure of work or extended hours of working.

2. Identify the people who might be affected by the engineering activity and how this might occur. This would include:

    - yourself, as you are carrying out the activity

    - other people working close to where you are.

3. Evaluate the risk and decide whether existing precautions are adequate, or whether more needs to be done. To evaluate the risk you should consider the likelihood of harm occurring and the possible severity of harm. Precautions are also known as control measures. Examples of these include safe systems of work, job rotation and the use of personal protective equipment.

    By inspecting work areas you can see **hardware control measures at the point of the hazards**. Look for guarding on machinery, extraction hoods in welding areas and electrical switches to control machinery. When assessing risk you will need to refer to **risk control systems**. These are the personal requirements such as the information, instruction and training so that you can carry out a task safely.

4. Record the findings from the risk assessment. This could be in a paper-based system or an electronic-based system. It ensures that the risk assessment can be referred to by everyone affected by the engineering activity.

5. Review the risk assessment and revise it if necessary. When you complete a risk assessment you should always enter a date for review. The review would be carried out after the activity is completed. This is because you will learn from experience and the review step allows you to reflect on what actually happened.

Key terms

Risk assessment – a five-step approach to making an activity as safe as possible.

Hardware control measure at the point of the hazard – examples: guarding, extraction hoods, electrical switching

Risk control systems – information, instruction, training, supervision to all employees and others affected by the engineering activity.

PLTS

By questioning your own and others' ideas during activities on the assessment of hazards and risks you develop your skills as a **creative thinker**.

Anticipating, taking and managing risks when handling materials and equipment in an engineering workplace in a safe and approved manner will develop your skills as a **self manager**.

By reviewing your progress during practical activities and acting on the outcomes you can develop your skills as a **reflective learner**.

Key term

Safe system of work – a step-by-step procedure to ensure a safe combination of the employee, the machinery and materials used and the level of training, instruction and supervision. Example: the safe and correct use of sharp saws and cutting tools.

Activity

Select an activity that you are about to carry out.

You are required to complete a risk assessment on the activity, using the five-step system described above.

You can refer to the case study on welding at the end of this unit.

Ensure that you make reference to recommendations on:

- the hazards of the activity
- the safety of materials and equipment handling
- the use of personal protective clothing.

Working at height

Falls from height are the most common cause of fatalities and serious injuries to employees. The main causes of falls are failures such as:

- not recognising the risk
- not using the correct access equipment
- not using a **safe system of work**
- a lack of awareness of one's own capabilities, or the capabilities of the equipment.

It is essential to understand that all work at height is hazardous. If at all possible, you should avoid having to work at height. You should ensure that safe systems of work are followed and always use the correct type of equipment.

Confined spaces

A confined space is any space that is enclosed where there is a specified risk of serious injury. It is easy to identify some enclosed spaces, such as deep excavations, storage tanks, or poorly ventilated rooms. Other confined spaces are not so easy, for example silos, furnace combustion chambers or vats.

Hazards and risks from working in confined spaces may already be present: for example, flammable gases, which could result in fire or explosion. They might flow into the confined space: for example, water from excessive rain or a broken water main might lead to drowning. Or they might be introduced into the confined space: for example, fumes generated while carrying out welding in the confined space.

Loss of consciousness might be caused by increased body temperature, owing to a lack of ventilation or to not wearing the correct PPE. It might also be caused by **asphyxiation** due, for example, to a build-up of gas, fume or vapour in the confined space, displacing the oxygen.

Entries into confined spaces are always hazardous. You will find that any work in confined spaces is strictly controlled by risk assessments, safe systems of work and **permits to work**.

Hot working

In welding, metals are heated to very high temperatures so that they can be joined together. The heat is generated by two basic methods. The first is by the use of acetylene mixed with oxygen and then ignited. This can also be used for cutting of metals. The second method uses electricity flowing through a welding rod and making contact with the pieces of metal that are to be joined.

Hazards associated with welding:

- burns – contact with hot pieces of metal.

- fire and explosions – by incorrect working techniques, welding too close to flammable materials, or in a flammable atmosphere.

- arc eye – when the UV light of the welding affects the eye, causing intense irritation.

- electricity is a hazard when electric arc welding – high current is being used to join metals together.

Forging of steels is a process where red hot material is reformed into another shape. An example are the hammers that you use in the workshop. These are hot forged to give them strength and to achieve the final shape without wasting too much material.

Hazards from hot forging:

- burns – contact with hot pieces of metal.

- noise – from the forging process, the steel billets being hammered into shape by the dropping of a heavy die.

- heat – from the forging process, can cause dehydration in the forge operators.

- vibration – into the arms and bodies of the forge operators.

Tools and equipment

When carrying out any engineering activity you must select the correct tools and work equipment for that activity. For instance, never try to use a screwdriver as a chisel. All tools and equipment must also be fit for use;

for example, hammers that you select for a particular job should be:

- the correct size

- the right type (e.g. carpenter's claw hammer, engineer's ball peen hammer)

- inspected: there should be no damage to the striking face, no splits or damage to the wooden shaft.

Dangers of not working to laid–down procedures

When you are employed in engineering, you must work to the laid-down procedures.

For instance, if you are asked to carry out a grinding activity, then you must wear the correct personal protective equipment. This would be shown in the risk assessment for that activity.

If you are performing a welding operation, you must follow the laid-down procedure for a safe system of work.

Similarly, the following laid-down procedure would apply to the use of a ladder for work at height.

- Decide whether you have to use a ladder, or whether you can use some other means of access.
- Select the correct type and length of ladder.
- Carry out a pre-use inspection of the ladder and workplace.
- Carry the ladder to site correctly, using the correct manual handling techniques.
- Use a two-person lift to place ladder in position.
- Place the ladder at the correct angle to the access point: 1 out and 4 up (see Figure 1.1).
- Ensure that the ladder extends 1 m above the access point.
- Make sure the ladder is suitably fixed before climbing on it.
- Use three points of contact when you are on the ladder: two feet and one hand.

1.1.4 Emergency procedures

Engineering workshop incidents

Everyone employed in engineering must know what they have to do in the event of an **accident** or **incident**. You will be informed of the process during your induction into an organisation.

Both accidents and incidents are warnings of failures in engineering activities. It is important that you as an employee report them to a superior. The main reason for reporting accidents and incidents is to prevent them happening again. Next time the outcome could be more serious.

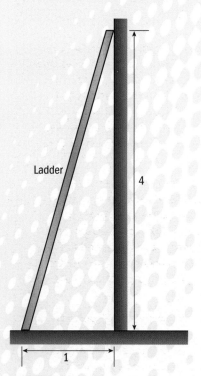

Ladder

4

1

Figure 1.1: The correct angle for a ladder.

Key terms

Accident – an unforeseen event causing injury or damage. Example: a person tripping over an uneven surface and breaking their arm.

Incident – an unforeseen event that narrowly misses causing injury or damage. Example: dropping a spanner off an overhead crane and just missing somebody.

Identification of appropriate qualified persons

In engineering organisations you will find that there are people who have additional responsibilities as well as their own jobs. These additional responsibilities might be as **first-aiders**, for example, or as **fire wardens**. Can you identify the first-aiders and fire wardens in your area of work? As well as being listed on a notice board, they quite often wear badges identifying their extra functions.

Actions in the event of an accident

On starting work at any organisation you will go through a process called induction. Part of the induction will be to instruct you on what to do in the case of an accident. Included in the instructions will be:

- make the area safe – by switching off machinery and making sure other persons are kept away from the accident or incident. Try to make the injured person comfortable, but only after assessing the situation.

- start the emergency plan that you have been instructed in. Call a supervisor, a colleague or press an alarm system.

- get a first-aider to help the injured person, their responsibility is to try to stabilise the condition of the injured person and summon help.

- in a fire emergency, turn your back on the fire and walk away from it. When walking away tell other persons of the fire situation. Sound the **emergency evacuation** signal by breaking a glass at a fire point, or by telling your supervisor of the fire. Make your way to the **assembly point** via the **escape route**.

Reporting routines

During your first few days of working in an engineering organisation you will be told about the reporting procedures you need to follow. The reason for the reporting is to ensure that any problems can be corrected as soon as possible. Examples of problems that need to reported are as follows:

- Hazards such as spilt oil must be cleaned up as soon as possible to prevent people slipping over.

- Malfunctions of machinery must be repaired so that operators are not injured when working the machine.

- Injuries must be dealt with by first-aiders as soon as possible to prevent the situation becoming worse.

- Near-miss occurrences must be dealt with to prevent possible injury next time.

Key terms

First-aider – someone who has been trained in first aid and is listed as having that extra responsibility. Their role is to deal with emergencies, to try to preserve life and prevent situations worsening until medical help arrives. They also treat minor injuries such as cuts and abrasions.

Fire warden – someone trained to ensure that everyone evacuates a building during an emergency evacuation. Their role is to ensure that the evacuation is carried out in an orderly manner and to check the building for any persons who have not responded.

Emergency evacuation – process used during a fire, gas release or other emergency to ensure that everyone affected leaves the building to a place of safety, usually outside and designated as an assembly point. The evacuation is initiated by the sounding of an alarm such as a two-tone siren.

Escape route – used during evacuation. Designated by green and white signs, showing a white arrow to indicate the direction, a walking person and sometimes a door. This sign category is a safe condition because it shows the route to a safe place.

Assembly point – a place of safety; a signposted position outside a building, usually in a car park or on a sports field.

PLTS

By questioning your own and others' ideas during activities on the assessment of hazards and risks you develop your skills as a **creative thinker**.

By reviewing your progress during practical activities and acting on the outcomes you can develop your skills as a **reflective learner**.

Discussing your findings with other members of your group will help you develop your skills as a **team worker**.

Activity: Improving emergency procedures

Your senior engineer has asked you to make recommendations for improving the emergency procedures.

To achieve this you will have to investigate the existing emergency procedures for fire and first aid in your work area. Then you will have to decide after analysis whether improvements can be made.

Choose a workplace from school or college, such as a woodworking or engineering workshop. Just concentrate on fire and emergency evacuation and first aid cover.

Write a report on your findings.

BTEC Assessment activity 1: Preparing for and carrying out an engineering activity

(P1) (P2) (P3) (P5) (M1) (M3) (D1)

For this assessment you will need to think about an engineering activity that you have recently completed and produce a short report. The report should be in four sections.

- Images and brief statements showing how you handled materials and equipment safely.

- A list of bullet points stating how you selected and used personal protective equipment.

- 6 bullet points listing how you identified the hazards and risks.

- images and short statements describing how you prepared and carried out the activities.

(P1) (P2) (P3) (P5)

Grading tips

(P1) (P2) (P3) For the practical activity, you must use equipment safely, adhering to correct lifting and carrying techniques and following good housekeeping practices. A combined approach will be expected for the materials handling, equipment use, the explanation of roles and responsibilities and the identification of warning signs.

The keeping of a task sheet / logbook / diary will help you to be successful in the assessment

(P5) You should be proactive in your thoughts on the working environment and tasks. Think about the consequences of situations and actions. To achieve this criterion you are advised to carry out a basic risk assessment; this could be following an inspection of a workplace or using your work-placement experience.

To achieve (M1) you need to present evidence that you carried out a more detailed risk assessment of the activity. Identify four hazards and say how you would reduce the risk of injury.

(M3) When you have finished the activity, identify two or three ways that you could have improved it.

(D1) Prepare a safety policy for the work area that you have been in; this could take the form of a poster to be displayed in a workshop.

1.2 Be able to work efficiently and effectively in engineering

1.2.1 Types of engineering activity

Installing and commissioning equipment or systems

Installing equipment involves fixing or placing of machinery or work equipment into position ready for use. Examples would be the permanent installation of a material-cutting machine such as a lathe or a drilling machine. The machines have to be placed in the correct position, then bolted to the floor and connected to the electrical system via an isolating switch. This will ensure that the machine is safe for a person to operate it. You will probably use these types of machinery for project work.

Examples of installing systems might be building and fixing a fluid power system for an automated assembly machine, or an electrical system to supply power to engineering machinery in a workshop.

Commissioning follows the installation of machinery or electrical systems. It ensures that the installation procedures have been followed and that every machine part works and is safe to operate. At this time any adjustments can be made to ensure efficient and safe working.

Machining and manufacture of products or components

Many products have to be machined. Take the engine of a car, for example. All the individual parts have to be machined so that they can be fitted together accurately and then work effectively as an engine unit.

Machining involves two basic processes:

- removal of material, such as using a **lathe** to reduce the diameter of a shaft, drilling a hole in a piece of aluminium so that a bolt can pass through, or sharpening a chisel.
- reshaping of material, such as **plastic injection moulding** of components, **forging** a spanner, or **drawing copper wire**.

The term 'manufacturing' applies to a vast range of products and components. Manufacturing generally involves several machining processes. An example would be an electric light bulb. This requires the individual machining or reshaping of the individual parts of the bulb and then the final assembly and packaging.

Key terms

Lathe – a machine that grips a piece of material and rotates it while tools of various types are applied to remove material.

Plastic injection moulding – process in which plastic pellets are fed into the cavity in a hot moulding tool to form products such as plastic spoons.

Forging – hot-working process in which red-hot pieces of steel are placed in the cavity of a forging tool and then shaped – into a spanner, for example.

Drawing copper wire – copper is very ductile and can be progressively reduced in diameter, by drawing it through dies, to various sizes for wires and cables.

Some typical machined components

Key terms

Servicing – routine work on machinery. Examples: oiling and greasing moving parts to prevent wear; changing filters that are partially blocked.

Maintenance – routine activities carried out to prevent or reduce machinery breaking down. Breakdowns cause production stoppages.

Maintenance personnel – mechanical, electrical and fluid power specialists such as engineers, planners and technicians.

Servicing and maintenance of plant and equipment

All work equipment must be serviced and maintained. For example, you will probably have carried out some **servicing** or **maintenance** on your bicycle at home. This might have involved oiling the chain, repairing a puncture in a tyre, or cleaning the bicycle after using it in muddy conditions.

These railway maintenance personnel are wearing brightly coloured PPE, for added visibility and safety

Servicing and maintaining work equipment follows the same basic servicing or maintenance on your bicycle, but on a larger scale. Machinery, work equipment and hand tools must be maintained in a safe, healthy and efficient condition. Engineering organisations employ specialist engineers, called **maintenance personnel**, who use a system called planned preventive maintenance to ensure that machinery and equipment works efficiently.

Prepare the work environment

When you are asked to carry out any engineering activity you must follow the general rules for preparing the work area:

- The area must be free from hazards. Ensure that there is enough space to work by removing any unnecessary items, bins or materials from the work area.

- Safety procedures must be followed. Engineers carrying out any work activity must be familiar with the risk assessments, safe systems of work and material details. This should ensure that all the hazards and control measures have been accounted for.

- PPE must be correctly selected and worn for the activity. For example, you would need to wear goggles and a boiler suit when operating a grinding machine.

- The tools selected must be appropriate for the activity. You would be expected to carry out a **pre-use inspection** or check all tools before use to ensure that they are safe and in a usable condition. For example, hammers should not have split wooden handles and the striking faces must not be damaged; spanners must be the correct size and the jaws must not be spread.

Preparing for the activity

When you are asked to carry out any engineering activity, your preparation should follow a general set of rules:

- You must obtain all necessary drawings, plans, specifications and instructions before you start work. These give you the details and information on the work to be carried out that you need to complete the activity.

- You must have authorisation to carry out the activity. This means that you are trained and experienced and are instructed by a manager or supervisor.

- You must obtain the correct materials and components: for example, raw materials for machining or processing components, or the correct oils, greases, wipes and replacement parts for maintenance activities.

> ## Key term
>
> **Pre-use inspection** – term used for the inspection and checking of tools or work equipment before use. It is good practice before starting an engineering activity.

- You must ensure that raw materials, finished products, tools and work equipment are stored correctly. Remember that good housekeeping ensures high levels of health and safety. There should be a place for everything and everything should be in place.

Completing the work activity

On finishing the engineering activity you will be expected to:

- have completed the task to the drawings, specifications and instructions given.

- complete any documents that are required to show that the task is finished: for example, a job card for completion of an engineering task, or a record card for completion of a maintenance activity. These documents can either be hard copy or input to a terminal.

- have returned all drawings, work instructions and tools to their correct storage places.

- have disposed of all waste to the correct storage areas: for example, unusable tools, swarf, material offcuts and replacement parts to a skip and waste oils, greases and wipes to a flameproof container with a lid.

1.2.2 Working relationships

Contributing to organisational issues

It is important that everyone involved in engineering activities works as a team. When you are working, you will find that you contribute to the team effort. Reliance on each team member is important, not only to complete the task, but also to ensure that it is completed in a safe and healthy manner.

Improvements in work practices and methods

As you complete tasks and activities you will find that you learn and gain experience. This will help you carry out similar activities in the future; it can be said that you are becoming competent. You will find that this is often referred to as a personal learning curve.

Competence means that you will choose the correct tools, materials and information for a task or activity.

Other improvements would be in the **quality** of work carried out by you as an experienced worker.

Improvements in the quality of finished products or services result in customer satisfaction. Customer satisfaction is important to engineering organisations, because it leads to further or increased orders.

Internal communications

Internal communications are an important function within engineering activities. For example:

- Operators of machinery must be instructed in the work they have to complete. For example, the communication might be a **works order** issued by the sales office.

- When a machine is to be maintained, the machine manufacturer's schedule must be consulted. This will describe how often maintenance has to be carried out and to which parts of the machine.

Other internal communications are risk assessments, safe systems of work, permits to work and **policies and procedures** for health. These will become familiar to you as you progress through this course and in the working environment.

Activity: Servicing a car

Ask someone who owns a car if you can carry out an analysis of their manufacturer's service book. Then answer the following questions;

1. Who does the car owner communicate with when they take the car to the garage?
2. How do you think the instructions are passed to the mechanic who will carry out the service to the car?
3. What would the working relationship be between the garage owner and the mechanic?
4. How often does the car have to be serviced?
5. List the items that are replaced at the service intervals
6. Are there other items on the car that may have to replaced outside the normal service?
7. Is the servicing requirement given in miles or time?
8. List the disposable items that have to be replaced during the service.

Dealing with problems affecting engineering processes

Despite careful planning before engineering activities take place, you will find that problems can occur. When you carry out your practical work you should try to foresee any possible problems.

You will discover that there are many reasons why problems can occur. For example:

- There might be a shortage of materials. The material requirements for any process should be ordered, or should be available, before the start. If they are not available, then production will be affected.

- The necessary tools and equipment must be ready to use and be listed on the works order before the start. All activities should be pre-planned so that the tools and equipment are ready to use.

- Engineering drawings, plans and specifications must detail all the information that you need to perform the activity.

- Meeting the quality requirements can sometimes be a problem. Quality is determined from the drawings and the specification. A high-quality product or service is going to cost more. You will find that there is a fine balance between the time spent on the activity or process and the final cost.

- You will find that people involved in the process can also be a problem. This can be due to lack of competence, to rushing the work by taking short cuts, or to not working effectively and efficiently.

PLTS

Anticipating, taking and managing risks when handling materials and equipment in an engineering workplace in a safe and approved manner will develop your skills as a **self manager**.

By reviewing your progress during practical activities and acting on the outcomes you can develop your skills as a **reflective learner**.

Activity: Improving workflow

You have been asked by your manager how the flow of work through an engineering organisation could be improved.

Select an engineering activity that you are familiar with. Examples could be maintenance of light fittings, or the production of injection-moulded plastic parts.

Then, using the list above as a guide, identify how the activity could be improved.

Working with others

Teamwork is an important part of many engineering activities.

Designers work with production engineers to ensure that the products they design can actually be made. Mechanical fitters and electricians work together as a team to maintain machinery.

Teamworking results in efficiency, consistency and a safe and healthy working environment for everyone involved.

You will find that you are more comfortable working in a team that you are familiar with. Problems can occur when new people have to enter into the team. Unfamiliarity can lead to a lack of confidence in each other when the existing team members are unsure how the newcomers will react in certain circumstances.

Management and supervision are important: this is where the direction and the information for the team flow from. Direction and information are necessary in all areas of engineering so the objectives for all activities and processes can be achieved.

As you progress in engineering you will find that management is built on trust and respect from all persons.

External relationships

Customers are essential for any engineering organisation. Without them, there would be no business. In any contact with customers you must be courteous and listen to their needs.

Suppliers are the companies that supply the raw materials, finished products, cleaning and office items, or any other products. These are required so that engineering activities can be carried out on your site.

Engineering companies use contractors to complete activities if they lack staff with the appropriate skills. Examples of contracted activities might be:

- building work
- window cleaning
- electrical work
- painting and decorating
- catering.

Activity: Improving working relationships

Imagine that you are working for a double-glazing company that manufactures and fits windows and doors in domestic properties. Your role is as a salesman and fitter.

Discuss in small groups the working relationships between yourselves as salespersons and fitters with the following persons:

1. your customers
2. your suppliers
3. the use of contractors.

Think about how you would treat the customer and what would you have to offer them?

What would be the service that you were offering to the customer? Are there other general issues that you would have to consider in the relationship, such as working safely when on their premises.

After your discussions write a short statement on the relationships, the services provided and any other general requirements in each case.

Can you identify strengths and weaknesses (areas for improvement) in the working relationships?

PLTS

By questioning your own and others' ideas during activities on the assessment of hazards and risks you develop your skills as a **creative thinker**.

Anticipating, taking and managing risks when handling materials and equipment in an engineering workplace in a safe and approved manner will develop your skills as a **self manager**.

By reviewing your progress during practical activities and acting on the outcomes you can develop your skills as a **reflective learner**.

Discussing your findings with other members of your group will help you develop your skills as a **team worker**.

Assessment activity 2: Emergency procedures and working relationships.

Suppose you have been carrying out a task in a workshop and an emergency occurs for example:

- a fire breaks out
- a person is injured
- there is a chemical spillage
- someone hurts their back lifting a heavy object.

P4 Describe the emergency procedure which would have to be followed after an incident of this type- you only need to consider one specific incident.

1. How would you report the incident?
2. Consider the documentation you would use?
3. Who would you make a report to about the incident?
4. If someone is injured how would you get help?

P6 Produce a brief overview of how good working relationships can be maintained with colleagues giving three examples.

M2 Have a look at the emergency procedures that are laid down in your workshop. Pick out the three improvements that you think would make them more effective.

D2 Write a short report in which you reflect on how you interact with colleagues when carrying out practical work. Are there any things you or a colleague could do to improve the working relationship?

Your tutor will observe the preparation and completion of a task.

Your tutor will observe the preparation for and completion of the assessment.

Grading tips

P4 A combined approach is required so you will be expected to research the reporting procedures after an emergency. You can undertake activities required for the completion of the assessment in an area that is familiar area to you, or in an unfamiliar area

P6 When working with other people it is important that you have a good working relationship with them. Ensure you obtain a witness statement from your tutor detailing how you have avoided conflict and worked well as a team.

M2 List in order of importance and ensure they include a description of how and why they make an improvement. Who would you present these recommendations to?

D2 To support your written evidence, include examples of real situations that you have been in.

 PLTS

By exploring hazards and risks associated with an engineering activity you will develop your skills as an **independent enquirer**.

 Functional skills

By writing a report to describe the emergency procedures to be followed in response to an incident, you will develop your **writing** skills.

WorkSpace James Hammond

Risk assessment engineer

James receives a work order requirement for a 200 mm diameter pipe to be fitted between two water vessels.

The senior engineer has requested that a risk assessment be completed before the start of the activity.

As the engineer responsible for completing the risk assessment, James has extensive knowledge of the water process and of the welding of pipework.

Using the standard risk assessment format, James must analyse all the foreseeable hazards and associated risks.

The analysis is broken down into steps, as follows, to ensure that the welding activity is carried out in a safe and healthy manner.

1. Collect all the information for the activity. Plans, drawings and specifications of the work area would be required, followed by a site visit. The site visit gives the local information.

2. Involve the welder who has to carry out the welding operation in the risk assessment process. This ensures that their expertise is used.

3. Identify the hazards: fire, explosion, heat, welding sparks and fumes.

4. Determine who is at risk and how they could be harmed – the welder and other persons in the vicinity.

5. Evaluate the existing precautions and control measures and see whether they could be improved. Existing precautions: a competent person, wearing the correct PPE and using the correct welding equipment. Improvement: correct choice of fire extinguisher nearby.

6. Record the assessment on the standard format paperwork, making sure that everyone involved is given a copy.

7. Carry out the welding activity to connect the pipe into the water system.

8. Review the risk assessment once the welding activity is completed.

Think about it!

1. Why would a risk assessment be required for this activity?

2. What does competence mean?

3. Why should the welder be involved in completing the risk assessment?

4. Can you pick out the standard five steps to risk assessment in the steps listed above?

Just checking

1. What is a hazard?

2. Explain what is meant by risk.

3. Define an accident.

4. Define a near miss.

5. List the five steps used when completing a risk assessment.

6. List the three reasons why safe working is important in an engineering workplace.

7. List three things that the Workplace (Health, Safety and Welfare) Regulations require an employer to provide for their employees.

8. Define the meaning of manual handling.

9. Give an engineering activity that is considered to be hot working.

10. What is your role as a team member?

11. What are your instructions on waste materials.

Assignment tips

- Always follow instructions.

- If you are not sure, then ask somebody who knows.

- Be aware of the hazards and the risks from any engineering activity.

- Be familiar with risks assessments and safe operating procedures.

- Always wear the correct PPE for the activity.

- Ensure that you keep up with the paperwork requirements.

- Carry out a risk assessment for an activity.

- If an accident or incident occurs, report it.

- Ensure that you know the emergency evacuation route.

- Never use a machine unless you have been trained and feel confident to use it.

2 Interpreting and using engineering information

When engineers present technical information, they should do it in a standardised way so as to prevent any ambiguity. Cutting metal and building products is expensive; it is even more so if people make mistakes, and produce scrap. This can happen if they are not given clear instructions about what to do.

A really important skill that you will use when working as an engineer is being able to access and use technical information. The types of information that you will be working with might be data taken from technical drawings, documents, sketches, circuit diagrams, job sheets or manuals.

Suppose you are shown a drawing of a component that is to have a length of 12. Now, is this measurement in millimetres, or should it be in inches? How do you find out? That's what this unit is all about: extracting engineering information and using it correctly.

The unit starts by investigating different types of engineering information, and ways of presenting it. This is followed by a look at how the drawings and documentation used in workshops are stored, catalogued, and kept up to date.

Learning outcomes

After completing this unit you should:

1. know how to interpret drawings and related documentation

2. be able to use information from drawings and related documentation.

Assessment and grading criteria

This table shows you what you must do in order to achieve a **pass**, **merit** or **distinction** and where you can find activities in this book to help you.

To achieve a **pass** grade the evidence must show that you are able to:	To achieve a **merit** grade the evidence must show that, in addition to the pass criteria, you are able to:	To achieve a **distinction** grade the evidence must show that, in addition to the pass and merit criteria, you are able to:
P1 extract information from engineering drawings and related documentation to enable a given task to be carried out **Assessment activity 2.1 page 39**	**M1** identify gaps or deficiencies in the information obtained that need to be resolved to enable a given task to be carried out **Assessment activity 2.1 page 39**	**D1** justify valid solutions to meet identified gaps or deficiencies with the information obtained. **Assessment activities 2.1 page 39 and 2.2 page 48**
P2 select and use other information sources to support and check information provided **Assessment activity 2.1 page 39**	**M2** identify improvements in the care and control procedures used for drawings and related documentation. **Assessment activity 2.2 page 48**	
P3 identify and obtain relevant drawings and related documentation to carry out and check own work output **Assessment activity 2.2 page 48**		
P4 complete all necessary production documentation related to own work output **Assessment activity 2.2 page 48**		
P5 describe the care and control procedures for the drawings and related documentation used when carrying out and checking own work output. **Assessment activity 2.2 page 48**		

How you will be assessed

The type of evidence that you will be asked to present when you carry out an assignment could be in the form of:

- a short written report
- a tutor observation record
- an engineering drawing onto which you have written comments
- a checklist of data that you have extracted from data sheets and production documentation.

Ali, 17–year old engineering apprentice

This unit has helped me to appreciate just how important it is to communicate properly when working in engineering.

When I started on the unit I thought engineering communication just involved talking to people and working with drawings and documents. I now realise that the process is much more involved, because during my studies I also learnt how to extract data from manufacturers' manuals, read technical specifications, and work with reference charts.

There were several practical tasks and activities that involved finding out information when I was on work experience. One of them involved talking to a quality assurance technician about how their company tested products and recorded information into a database so that designers could check out the reliability of new products.

Over to you

- What parts of this unit might you find challenging?

- Which section of the unit are you most looking forward to?

- How do you think you might prepare for the unit assessments?

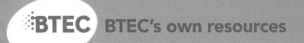

2.1 Know how to interpret drawings and related documentation

Start up

Getting the job done

A flow chart is a diagram that maps out a sequence of events so that it can be easily communicated to other people. Flow charts are widely used in industry.

Working with a colleague, find an example of a flow chart in a service manual or on the web. Identify places in the chart where yes/no decisions occur.

List what happens when a car is left at a garage to have two front tyres replaced. Draw a flow chart that shows the sequence of operations. Show the chart to a different colleague and ask them to talk you through it. If they were a mechanic, new to the job, would they be able to follow your instructions and safely change the tyres?

When you list the operations, think about safety checks and yes/no decisions: for example, are the tyre pressures correct?

A bit more about tyre pressures: there are two places where you can find the correct values for the car. Check with your tutor that you have identified them successfully.

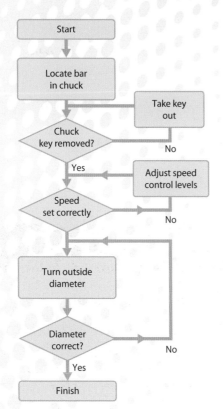

Figure 2.1 An example of a flow chart, for turning a diameter on a lathe.

2.1.1 Information

Engineers work with many types of information when they are designing, manufacturing, commissioning and maintaining products and services.

Suppose you are a trained technician based in a college workshop, and a learner asks you to help them with their manufacturing project. They are experiencing problems machining a fairly complicated component that they have designed and the tutor suggests they pass the job over to you. Before getting started, you need information and the absolute minimum would be the quantity to be made, the physical dimensions of the component, and the material to make it from. Going a step deeper, you will also need to know about the accuracy of dimensions, how the surfaces of the component are to be finished off and whether it should be heat-treated. Can you think of any more information you might need?

Engineering information can be placed into several broad categories:

- assembly instructions for fitting parts together so that they are accurately located and correctly **orientated**

- characteristics that are checked after a circuit has been connected up: for example, the current flowing in an electric circuit, or the rotational speed of a hydraulic motor when the pressure supply is turned on

- dimensional detail such as units of measurement, physical sizes, **tolerance** and **surface texture**

- manufacturing detail such as machining processes, tooling, and assembly sequences

- symbols and abbreviations that are used in drawings and documentation. British and European standards specify which ones should be used for particular applications, for example BS 499-1:2009 for welding applications.

2.1.2 Engineering drawings

Working drawings can be produced by hand using paper and pencil, or in a more sophisticated way with a CAD system. When someone looks at a drawing they are performing a process called reading the drawing. This involves looking at images, text, numbers, symbols and abbreviations, interpreting what it all means, and then deciding what to do with this information. Anyone who is involved with the technical aspects of engineering must be able to read drawings.

What types of drawing do engineers work with? Think about a domestic appliance such as a washing machine. It's made up from lots of components that are fitted together to make the finished product, which is referred to as an assembly. The machine's drive motor is also an assembly of parts, but is called a sub-assembly, because it contributes to the finished product. It's quite possible that the motor is manufactured in a completely different factory, which just specialises in motors.

The person who buys the finished machine needs only a drawing that shows its outside dimensions and connection details. The person machining the spindle of the drive motor needs a very detailed drawing, containing lots of manufacturing information.

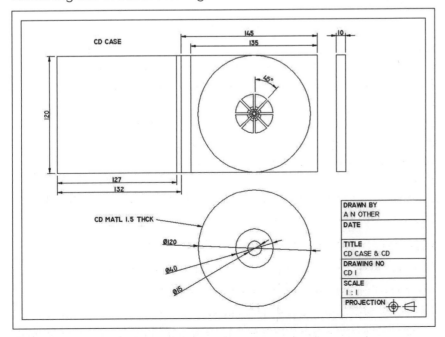

Figure 2.2 A combined component and assembly drawing

Did you know?

The BSI (British Standards Institution) is recognised globally as a producer of standards and information products that promote and share best practice in industry and commerce. You can access the catalogue of over 27,000 British Standards at the BSI website, see the hotlinks section on page ii, and a search facility allows you to find the one you need.

Activity: Standards

Working with a partner, identify two reasons why standards are used.

What is the BS number of the standard that someone would refer to when designing an electronic circuit?

Think about the special names we give to the different types of drawing used by engineers:

- component – this contains all the information needed to manufacture a component

- sub-assembly – this shows how the components of a sub-assembly fit together, and includes a numbered list of components (called a parts list), and assembly instructions such as torque-tightening values for bolts

- general assembly – this shows how everything fits together to make the finished product – that is, components and sub-assemblies. It includes a parts list, assembly instructions, and the dates when any design modifications were made

- fabrication assembly – this is similar to a general assembly drawing, but is specific to structures made from sheet materials that are joined by welds, adhesives or mechanical fixings

- repair/modification drawing and notes – used by maintenance technicians

- installation – gives information about the overall dimensions of a product and how to install it: for example, the instructions that are provided with a bathroom extractor fan

- circuit diagrams – the most common ones are wiring, electrical, hydraulic and pneumatic. They show how components should be linked together, and also provide information about the operation of the circuit, for example test-point voltages or pressures.

Remember

BS 8888 is the standard that you should refer to when creating an engineering drawing.

Activity: Reading drawings

Work in pairs or small groups.

Find an example of each of the different drawing types listed above.

For each drawing, find two specific pieces of information: highlight these with a marker pen, and think about what the information means.

Set out your findings in a table like the one below.

Drawing type	Information	What it means
Component	1. M15 × 1.1	A 15 mm diameter thread with a pitch of 1.1 mm

PLTS

Reading and interpreting information presented in an engineering drawing will help develop your **independent enquirer** skills.

If you work with colleagues when reading and interpreting information from drawings and documents, you will be able to develop your **team worker** skills.

There are several other types of graphical representations that engineers use:

- Sketch – a relatively quick way of communicating information about a component or assembly, particularly if it is accompanied by notes. The most useful sketches are those that have dimensions in the correct proportions to the object that is being drawn.

- Schematic diagram – similar to a circuit diagram, but does not always have to feature standard symbols. An example of a schematic is a railway route map showing how cities and towns are linked. Schematic diagrams are a very user-friendly way of presenting complicated information.

- Flow chart – this shows how several individual actions are linked: for example, the sequence of events when a washing machine is switched on and it goes through a wash cycle. Flow charts use standard symbol shapes: the most common ones are an oval (start/stop), a rectangle (action/event) and a diamond (yes/no decision).

- Physical layout diagram – also called a plan: for example, a drawing showing where machines, benches and safety equipment are positioned in a workshop, and how the electrical and compressed air services are connected to them. This type of diagram is particularly useful when drawn with a CAD system, because it can be built up in layers drawn in different colours, such as black for machinery, blue for benches, and red for electrical supplies.

- Manufacturer's manual – hard copy (printed) or online (electronic) documents that contain exploded diagrams and images of products. A designer will refer to this type of manual when deciding which product to use for a particular situation: for example, when choosing the correct electric motor to fit to a machine for stamping out sheet steel components. Service technicians find this type of documentation very useful when ordering replacement parts.

Functional skills

Reading and extracting information from an engineering drawing will help you develop your skills in **English** (reading).

Key terms

Orthographic projection – the representation of a three-dimensional object drawn on a two-dimensional surface using a series of linked views.

Pictorial sketch – a technical illustration that shows the faces of a three-dimensional object in a single view drawn on a two-dimensional surface.

Activity: Communicating information

Find a simple component that has flat faces and a hole or a slot. Measure it up and make a **pictorial sketch**. Now turn this information into a proper component drawing by sketching out the views in **orthographic projection**.

Add dimensions and any other information you think is important.

Arrange with your tutor for someone to manufacture the component you have just drawn. Compare the new component with the original one:

- Are they the same?

- Did you interpret correctly what you saw and measured?

- Did the person in the workshop interpret your drawing correctly?

- If things went wrong, what needs to be done to prevent this happening again?

PLTS

Arranging to have a component manufactured and then checking that it has been produced to specification will help develop your **self-manager** skills.

2.1.3 Related documentation

Engineers have to follow laid-down procedures when carrying out a task. For example, suppose you are a maintenance technician about to service a conveyor belt in a factory that assembles washing machines. You have to get underneath it to replace some rollers and a motor.

What do you need to know before starting the job?

Obvious things come to mind. Which tools and equipment should I use? Have I got the correct replacement parts? How do I isolate the electrical supply? Are there special health and safety requirements when servicing conveyor belts?

Depending on how you get paid, you may need to fill out a time sheet, but there certainly will be a job completion and test certificate to sign when you have finished.

The paperwork involved with carrying out a job is called documentation, and it can be in **hard copy** or **e-format**.

One category of documentation relates to work instructions, and these might be:

* Operation sheet/job card – this sets out a sequence of events. For example, to manufacture a simple component it might say: mark out, cut to length, drill hole, etc. The operation sheet will also state what tools and equipment are needed to do the job.

* Test schedule – this has all the necessary detail for someone to be able to carry out a test procedure on a system or piece of equipment. It will contain numerical data relating to measurements that someone may make, such as voltages, pressures or frictional resistance.

* Manufacturer's manual – this contains full instructions relating to the assembly, testing, installation and servicing of equipment.

* Weld procedure specifications – welding is a specialist process, and the quality of the joint will be specified by a British Standard in situations where joint failure would have serious consequences. There is a lot of documentation about welding because of all the ways it can be done, for example by oxy-gas or **TIG**.

Key terms

Hard copy – information printed on paper.

e-format – information displayed on a computer screen.

TIG – tungsten inert gas welding. The weld area is shielded by an inert gas, such as argon, to prevent atmospheric contamination of the hot surfaces.

Activity: Welded structures

Work in pairs or small groups. Identify three products that are welded fabrications: for example, a pair of car ramps for DIY use, the air receiver tank of a portable compressor, or a bicycle frame.

For each product, discuss the consequences of weld failure.

Investigate where you can access weld procedure specifications.

Investigate what is meant by non-destructive testing (NDT) of welds.

Another category of documentation relates to quality control information:

- Standards – these can be UK, EU, international, or linked to a specific type of product or process. Almost every product or service that people use is covered by a standard. Standards are there to protect the customer from bad service. For example, most people buying a quality plasma TV would be very annoyed if it failed after just three months' service.

- Reference tables and charts – when products are manufactured, various quality control checks will be carried out. The data generated is stored, and quite often is used for statistical analysis. A quality assurance technician might compare measured data with reference values held in a table. For example, suppose an automated machine is set-up to machine the length of a component to a given dimension. If the machine turns out thousands of components an hour, it is not economic to check every one, and so sampling is carried out. At set intervals a component will be checked, and its measurements tabulated and charted. The technician will assess what is happening, and if the dimension starts to go out of tolerance they will ask for the machine to be stopped and adjusted.

2.1.4 Tasks

Suppose you are working in industry as a technician and at the start of each week you are given a schedule of jobs to carry out. These are tasks which need to be completed to an agreed standard. Some examples are manufacturing or modifying a product, installing or repairing a piece of equipment, or planning how to carry out maintenance on a system. You may well have all the drawings and documentation that you need, but there can be other specialised information that you will need to be referring to.

This can be very specialised. It might be:

- Connection configurations – for example the pin layouts of electronic components, or the dimensions of the threaded connector on the end of a high-pressure hydraulic hose. In today's global environment, parts and assemblies have to be interchangeable worldwide. It doesn't matter where you buy a cable fitted with **USB**-A plugs; you know it will fit your computer because the connector has been produced to an international standard.

- Reference charts for limits and fits – parts that fit together have varying degrees of fit. For example, a shaft running in a plain bearing will have a clearance fit, so that it can turn easily; a peg hammered into a hole will have an interference fit, so that it does not drop out. BS 4500 is the UK standard covering limits and fits, and engineers refer to its reference chart when specifying the dimensions for parts that need to fit together.

- Tapping drill reference charts – lots of components have holes that are screw-threaded so that fixing bolts can be screwed into them.

> **Key term**
>
> USB – universal serial bus: a standard specification for connecting computers and peripheral devices. Currently there are over 6 billion USB devices being used around the world.

It is very important to drill the correct diameter hole before cutting the thread. If the hole is too small, the threading tool will jam up and snap; if the hole is too big the thread will be weak, and may strip out.

- Bend allowances – before bending a piece of sheet metal it has to be marked out and extra length added. This is called the bend allowance, and will depend on the metal thickness, bend radius and angle of the bend. Values for bend allowance are tabulated, or you can search for a bend calculator package on the web.

- Metal specifications – design engineers need to access data about the properties of metals, for example their strength and corrosion resistance. Also, they will want to know where they can buy raw materials. This type of information can be found on the web, in reference tables, and in stockholders' catalogues.

- Manufacturers' data – the range here is huge, and will depend on what you are working on. You might want know about types of welding rod, and how best to use them; or perhaps what type of adhesive to use to stick two materials together; or perhaps what paints and surface finishes are available. Manufacturers' data is often presented in two parts: the available products, and how to work with them, such as advice on surface preparation, thickness of adhesive, curing times and health and safety. This type of data can still be found in paper-based format, but increasingly manufacturers are putting it on their websites as downloadable PDF files.

Did you know?

PDF stands for portable document format. A PDF file is a very efficient way of sending large amounts of data electronically. It will take up much less memory space in a computer than a word-processed document containing the same numbers of words and images.

Key terms

HDMI – high-definition multimedia interface: a compact audio/video interface for transmitting uncompressed digital data. Set to become the next big seller in connection technology.

Machining centre – a multi-function machine on which complex milling, drilling and boring operations are performed.

Quality control – methods and procedures for measuring, recording and maintaining quality targets.

Activity: Connection configurations

Work with a colleague. Find the pin configuration for an **HDMI** cable.

Identify the British Standard that covers the dimensions of hydraulic connectors and adaptors.

Identify the correct tapping drill to use for an M20 × 1.5 threaded hole.

Case study: Dave, design engineer

Dave works in the design department of a company that manufactures pipe fittings and valves for use on oil rig platforms. The business has a very sophisticated computer-aided design (CAD) system, which is linked to computer numerically controlled (CNC) machining centres.

Dave is a member of the rapid response team, which produces 'specials' for customers who require a fast turn-around time. He designs the component, selects a suitable material, and then passes data electronically to a **machining centre**. Here a technician sets up tooling, runs a simulation of the cutting sequence, loads raw material, and puts the machine into run mode.

Quality control data is then fed back to Dave's computer so that he can keep a check on what is happening in the machine shop.

Now answer these questions:

1. The CAD package that Dave uses has 2D and 3D options. What does this mean?

2. When choosing a material for a component, what type of reference sources will Dave use?

3. The components he designs have screw threads machined into them. Where will he look for information about standards relating to these threads?

Assessment activity 2.1

1. Ask your tutor for the name of a company (e.g. Triton or Mira) that manufactures electric showers for use in domestic bathrooms.

2. Find out how to buy spares online and then ask your tutor to select a specific model of shower on which a suitably qualified person could carry out a maintenance procedure.

3. Print off an assembly drawing, identify the can assembly, and make a note of the manufacturer's part number. **P1**

4. Fill in the online order form, print a copy and show it to your tutor. **P2**

5. Find information about fitting the replacement part and print this out. **M1**

6. A qualified person replaces a faulty can. When they run the shower it still does not seem to be working properly. They decide to do a full test against specification, but don't have all the information required, for example figures for flow rate and temperature stabilisation. How will they overcome this deficiency in information? **D1**

Grading tips

P1 To achieve this grade you must correctly identify the manufacturer's part number, order code and fitting instructions for the specified component.

P2 To achieve this grade you must present additional evidence that supports what you did for P1.

M1 The specification for the shower is not given in the spares section of the website. To find it you should have used another information source.

D1 If any of the information you needed was missing or presented incorrectly, describe how you got around the problem. You must be able to justify your actions. If you had no problems, ask your tutor for some technical information that is flawed, and identify solutions to overcome the deficiencies.

PLTS

Extracting information from an assembly drawing and using manufacturer's part numbers will help develop your **independent enquirer** skills.

Functional skills

Using an online database to find information about bathroom showers is a good way to develop your **ICT skills** in accessing, searching, selecting and evaluating data.

Did you know?

Components for the Airbus A380 aeroplane are manufactured in over 20 countries around the world. Final assembly is carried out at a factory in Southern France, and altogether thousands of people will have been involved in the production process. When the first prototype was put together, all the components fitted perfectly and on its first flight the aeroplane performed as expected. There are millions of components in an A380, each one with its own drawing and documentation that traces its history. That's a lot of information!

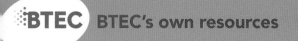

2.2 Be able to use information from drawings and related documentation

So how do we bring together everything from the first part of this unit and apply it to a particular manufacturing or engineering process that someone might be tasked to do? It's great to have information, but it's what you do with it that is more important.

Successful engineers always plan carefully before starting on a project. As the project progresses, they will refer to increasingly specific information.

Activity: It's your lucky day!

The company you work for has just taken delivery of a new HD-ready TV, hard drive/DVD digital recorder and video camera. The managing director (MD) wants to produce short in-house videos that showcase the business and the products it makes. The idea is to burn the videos to DVD so that they can be given to local schools before learners come for work experience.

The MD calls you into the training room and says, 'I can't get this lot to work, and I don't have the time to read the instruction manuals. I thought it was just a simple case of plug it together and away it goes. No such luck. I understand you're a bit of a "techie"; take as long as you like, but please get things to work for me!'

1. How would you set about getting things to work?
2. Make a checklist of the steps you will follow.
3. What specific information will you need?
4. Where will you find it?

2.2.1 Work output

Now think about the information needed when carrying out the following types of task:

- manufacturing a component using a machine tool such as a lathe

- assembling components into a finished product using hand and basic power tools, for example a soft-faced mallet and a power screwdriver

- designing a product such as an electronic circuit

- planning the maintenance procedure for a product, for example the routine servicing of a car engine

- carrying out maintenance on a product, for example cleaning up the C drive on a computer and updating its security software.

These are all tasks that require expert knowledge on the part of the person carrying them out. An expert will very often refer to information sources, because it can be quite difficult to hold all the information they need in their head.

Activity: Information sources

You are about to produce an external thread on a mild steel rod using a centre lathe and a single-point cutting tool. Where will you find information about setting the correct speed and feed rates?

PLTS

Finding technical information about electrical components or engineering materials will help develop your **independent enquirer** skills.

Case study: A manufacturing company

TDK-Lambda UK designs, develops and manufactures high-quality power supplies that are used in industrial and commercial applications. Many of these applications are 'mission critical'. This means that if a systems failure were to occur, the knock-on effect would be huge. For example, think about what might happen if the air traffic control system at an airport went down, or one of the major independent TV channels failed at peak viewing time.

TDK-Lambda's products are extremely well designed, and at all stages of their manufacture are rigorously tested against specification. Test results and other information are stored in a database so that there is full **traceability**.

Their range of standardised, modular power supplies is large, and information about products is presented on their website and in hard-copy brochures.

Now answer the following questions:

1. You are a design engineer working for a business that builds mobile TV studios for use at outdoor events such as the Glastonbury festival and the world snooker championship. You decide to use TDK-Lambda power supplies, but having looked at their catalogue you are not exactly sure which ones to specify. Whom do you need to talk to at TDK-Lambda?

2. What information will they ask you for?

3. How will this person communicate information back to you?

Key term

Traceability – the ability to trace parts and products back to their source.

Activity: Carrying out and checking own work output

Select a practical activity from one of the other units that you are studying on this First programme. For example, Unit 4 (Applied electrical and mechanical science for engineering) has several topics that involve performing experiments in a laboratory.

- Agree a suitable experiment with your tutor.

- Design a simple checklist that you can use to make sure you don't miss anything as you do the experiment.

- Read through the instructions for carrying out the experiment.

- Set up the equipment, obtain results, and process them.

- Complete your checklist and ask your tutor to sign it off so that you have evidence of checking your own work output.

2.2.2 Production documentation

Engineering factories use lots of different types of documentation to monitor the progress of products as they are manufactured. There are several different ways of presenting this documentation to a user. It might be as a printed form that someone fills in by hand, or it might be a display on a computer screen, with data entered via a keyboard or perhaps a hand-held wireless computer terminal.

Did you know?

Parcelforce and other delivery services use hand-held terminals to send tracking data back to a central computer system. A customer can check online the progress of a delivery.

A similar system is used in manufacturing industry when parts are moved around a factory before being fed to automated assembly lines. Components are either individually barcoded or placed in marked trays.

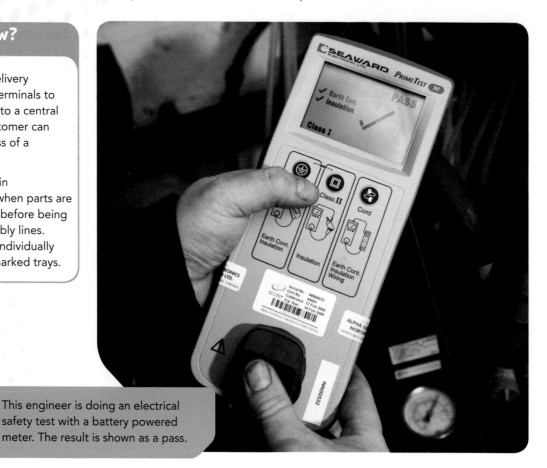

This engineer is doing an electrical safety test with a battery powered meter. The result is shown as a pass.

Look inside a new car, particularly under the carpets or in the boot. You will probably find small paper labels with barcodes on them. Sometimes there will be the signature or the initials of an inspector. When the car is delivered to the garage these will let the salesman check that it has been assembled correctly.

There will also be a colour code identification plate fixed to the chassis. If the car needs to be repainted after a crash the paint shop will know exactly which colour to use.

The car will also be delivered with a service record booklet, so that every time the vehicle is serviced this can be recorded. If the car is sold, the new owner can check to see if it has been properly looked after: the car is sold with a full service history (**FSH**).

Key term

FSH – full service history is a full record of all the servicing a car has received.

Test Specification for 555 Timer Circuit

Visual Inspection
Check printed circuit board for:
  No holes
  No bridges
  No burns
  No lifted tracks
  Chip orientation (pin one in the correct place)
  Resistors links and chip socket flat to pcb and
  with no gaps

Resistance tests (no power and with chip removed)
Pin 2 to Pin 4 on chip socket 2MΩ ± 10%
Pin 1 to Pin 8 on chip socket ∞
Pin 2 to Pin 7 on chip socket 1MΩ ± 10%
Pin 3 to Pin 6 on chip socket ∞
Pin 4 to Pin 5 on chip socket ∞

Voltage tests (5V supply with chip in place)
From Ground to Pin 8, 5V ± 0.25V
From Ground to Pin 7, 2.5V ± 0.25V

Figure 2.3 A typical test specification

 Did you know?

Someone selling a car with an FSH usually gets a better price for it than for one without this document.

Examples of production documentation you will come across in engineering are:

- Job cards, which track the progress of manufacturing processes or operations as they are carried out. Sometimes they are referred to as operations or planning sheets, but this is not strictly correct. A job card should be used to record the progress of a component or assembly as it moves through the production area of a factory. Originally job cards were literally pieces of card that could be signed off by production workers and inspectors at each stage of the

Works Order No:			
Part No: **Description:**		**Sales Order No:**	
Customer Name:		**Customer Standards Apply:**	
Customer Acc no:		**Certificate of conformity required:**	
Customer Ref:		**Check Issue:**	
Quantity:		**Check drawing:**	
Required by:			
Scheduled for:		**Issued by:**	
Scheduled completion:		**Signature:**	
Materials: **Req Qty**	**Item Code**	**Product description**	
Mat spec	**Issued Qty**	**Batch No**	**Initials**

A typical job card

Did you know?

If a product is not fit for purpose, then it will not meet the needs of a customer purchasing it. They will seek redress from whoever sold them the product, and this will cause extra expense for the manufacturer, because they will have to put things right.

Functional skills

Reading and understanding production documentation will help you develop your skills in **English** (reading).

Key terms

CIM – computer integrated manufacture. The entire manufacturing process is controlled by computer. This allows real-time information to be passed easily between departments involved with design, planning, purchasing of raw materials, manufacturing, quality assurance and other business functions.

Specification – a clearly stated requirement or detail relating to a product or system.

manufacturing process. They are particularly useful where traceability has to be maintained, for example in the aero industry.

Many businesses still use 'hard' job cards, but where computer-integrated manufacturing (**CIM**) systems operate you will find data is recorded via hand-held terminals and computer key pads.

- Test reports, which are used to record the results of tests carried out on components and assemblies. These might be very simple tests, such as measuring and recording the dimensions of a component, or complex ones involving test equipment. The purpose of a test report is to provide a written record of whether a product meets its design **specification**.

- Quality control documentation – used as part of the process to ensure that products are fit for purpose. This type of documentation relates to:

  - inspection, testing and checking of raw materials and bought-in components as they arrive at a factory

  - inspection and testing of manufactured components, assemblies and finished products

  - pass/reject rates for components and finished products

  - customer service – customer support, complaints and warranty issues.

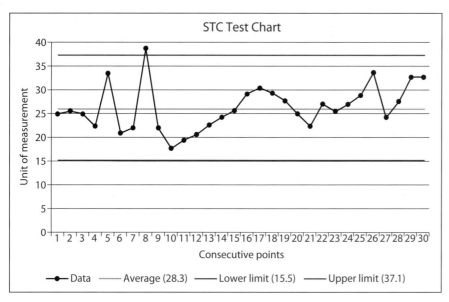

Figure 2.4 Typical quality control documentation: a test chart

2.2.3 Drawing and document care and control

Drawings and documents should be stored in ways that allow authorised people to access them easily. If someone has to spend time tracking down the drawing of a component, this will waste both time and money. Or think about what happens if there is a problem with a product that fails in service and someone has been asked to check whether it was correctly built and tested. They will need speedy access to quality control documentation.

What happens if a competitor gets hold of company documents and steals the ideas set out in them? Perhaps you did all the work set out in the documents, but they then profit from your efforts. You will be very annoyed, to say the least.

By setting up secure, properly organised storage systems, with controlled access and return procedures, a business can overcome these and other issues involved with keeping engineering data. The way this is done will be similar for paper and e-based systems.

To care for drawings and documents and control their use effectively, engineering businesses have procedures that deal with location and security. The following features are important:

- Storage – this might be electronic filing in a computer system server, traditional filing cabinets, hanging cabinets for drawings produced on paper and plastic film, or a microfiche system. A microfiche is a piece of photographic film that measures about 100 mm × 150 mm and contains printed information in a size too small to be seen by the naked eye. Documents and drawings are photographed and printed

Did you know?

The International Standards Organisation's ISO 9000 series of quality standards have a section that relates to the care and control of drawings and documentation.

onto the microfiche, which can be read using a very simple projection unit that does not involve the use of a computer. Many garages still use microfiches in their spares departments, because the system is very robust. It is also a reliable way of **archiving** information about cars that have gone out of production.

- Access and return systems that allow only authorised use of materials. Engineers working on computer systems have passwords set to a particular security level. People working with paper drawings book them in and out of a central store, in much the same way as borrowing a book from a public library.

- Reporting discrepancies is a very important issue. If the user of a document or drawing thinks that there are errors in what they are looking at, there will be a reporting procedure for their concerns. They record these concerns in writing and the information is passed back to whoever produced the original documents. If the problem is valid, then modifications are made to the documents or drawings so that they can be reissued. For traceability reasons it is important to have a system for logging these modifications.

- It is very important to having an identifiable link between all the documentation and drawings associated with a particular project. This is done using a top-down numbering system, which starts with the general assembly drawing and then works down through sub-assemblies, components and documents. It is very similar to organising folders and sub-folders in a computer system.

Activity: Storing documentation

Work with a colleague.

One of you types a full page of basic text, in font size 10 or 11 pt.

The other produces about half a page of similar-sized text, but then adds some images and colour effects.

Save both documents, then use the software 'properties' command to establish the size of each saved file.

How do they compare?

Did you know?

Approximately 15 000 pages of A4-size text-based documents can be stored on a 700 **MB CD-ROM**. In printed form the pages would fill a standard-size four-drawer filing cabinet.

At the design stage of a project, drawings are usually produced on-screen using a CAD system. They are then printed off so that people can discuss them and mark then up with any changes. These alterations are then incorporated in the stored drawing files and final versions are printed off. These hard-copy drawings are then passed to the workshop, where machine operators and technicians read them. This will involve physical handling, and possible folding for a large drawing that has to be pinned up next to a machine tool. Workshops can be dirty places, and so it is important that drawings and documents are kept clean, otherwise they may be difficult to read, and mistakes will occur.

Activity: Designing a numbering system

You and a colleague are working on the design and manufacture of a new mass-produced product. It is complex, and is made up from five sub-assemblies, each of which has at least 10 components. Various inspection checks are to be done during the manufacturing process, and the finished product will undergo testing to ensure that it conforms to its specification.

The product is designed to be updated every 3 or 4 years by making modifications to some of the components – particularly to a microprocessor sub-assembly.

As the product is new, its performance in service will be monitored by a customer services department.

All drawings and documentation will be stored electronically.

Design a numbering system that could link the various elements of the project.

PLTS

Completing production documentation after manufacturing a product, or setting up and testing a system against specification, will help you develop your **reflective learner** skills.

Functional skills

Writing about the care and control procedures for drawings and related documentation will help you develop your skill in **English** (writing).

People using paper drawings and documents should observe the following simple rules:

- Handle them carefully so that they do not get torn.
- Keep drawings and documents free of graffiti (however minor) by not writing on them or marking them up. This keeps them easy to read.
- Do not leave drawings lying around so that they get dirty.
- Fold drawings so that the title block is always at the bottom right-hand corner. This makes for easier sorting when thumbing through a pile of paperwork.
- If there is a problem with a drawing, ask for a new one, or seek clarification from the person who produced it.

Document control is a process that ensures that people are always working with up-to-date information. When a product or system has been designed and manufactured, there may be times when it needs to be updated, or perhaps modified if faults occur.

When modifications are made to drawings and documents, these are recorded in change notices, which are held at a central point.

Drawings and documents have issue numbers written on them. When a change is made, the date it took place is added, the issue number is updated and a confirmation signature (or initial) is added. This provides an audit trail, which can be used to back-track to the original documentation if the change proves to be flawed.

If any form of documentation, whether electronic or paper based, becomes damaged or missing, the problem must be recorded in the document control system and recovery procedures must be invoked.

BTEC

Assessment activity 2.2: Working with production documentation

This assessment activity is best linked to a unit that involves you being in a workshop, for example the maintenance or manufacturing department of a factory or school/college.

1. Agree with your tutor a practical activity to be carried out. This can be done individually, or with a colleague.

2. Get hold of drawings, planning sheets, parts lists or any other documentation that you will need to refer to. **P3**

3. Design a job card on which you can record your progress as you carry out the practical activity. If you are machining something, then you will need to record inspection data, such as dimensional measurements. If you are carrying out a maintenance procedure, you may need to record which components have been replaced, and the results of testing. **P4**

4. Carry out the practical activity and fill out your document as you go.

5. When you have completed the activity, ask your tutor to role-play a quality control technician, and see whether they will sign off your job card. The idea here is to confirm that everything you did was to specification.

6. Now think about where you obtained the documents, and what you did with them when you had finished the job. Did you have to sign for them? How well did you look after them? **P5**

7. The company you work for is reviewing the operation of its document control procedures. The current system is very paper based, and the manufacturing director has decided to put in a completely new e-based system designed in-house. The MD asks that you and a colleague produce a design brief for the proposed new system. **M2**

8. Review your success in carrying out tasks 1–7. **D1**

Grading tips

P3 Before starting on the practical activity, you should confirm with your tutor that you have correctly selected all the necessary drawings and documentation.

P4 As you progress through the practical activity, update your job card. Make sure that it is fully completed when you finish the task. What will you do if the 'quality control technician' refuses to sign off your job card?

P5 To achieve this grade you should describe the procedures for access and return to storage procedures for the drawings and documentation that you used. You must also describe how you looked after them when in your possession.

M2 To achieve this grade you should write about 250 words to cover task 7 (above). It will improve your evidence if you include a flow chart.

D1 If any of the drawings and documents had information that was missing, or which was presented incorrectly, describe how you got around the problem. You must be able to justify your actions. If you had no problems, ask your tutor for some technical information that is flawed, and identify solutions to overcome the deficiencies. For example, suppose there are missing dimensions on a drawing: you might find them by talking to the person who produced it.

PLTS

Evaluating how you used documentation to come up with proposals for a new document control procedure will help develop your **reflective learner** skills.

Functional skills

Designing and completing a job card will help you develop your **writing** skills.

Emma

Applications engineer

HiTeb Ltd is a small business that manufactures instrumentation systems. A Formula One motor racing team has fitted HiTeb's equipment to a new car they are developing for next season's Grand Prix series. As the car is test driven around a track, load sensors measure the forces in the suspension and other highly stressed components.

Data is streamed back by telemetry to the team's design headquarters, where it is processed by a computer system. Engineers then compare their design calculations with the measured values of the forces and make adjustments to the set-up of the car's suspension.

Emma is an applications engineer working for HiTeb. She has been called to a meeting with the chief designer of the F1 team, because there is a problem. Some of the 40 sensors fitted to the car seem to be malfunctioning and a significant amount of the data coming through does not agree with the design values.

Emma puts together an action plan to resolve the problem:

- Have the correct sensors been selected from HiTeb's catalogue?
- Have the fitting instructions been correctly followed?
- Are the sensors faulty?
- Is there a problem with the telemetry system?
- Is there a problem with the software that processes the readings from the sensors?
- Are the original design calculations correct?

Think about it!

1. What information is needed when selecting a sensor?
2. Where could a technician find information about fitting a sensor to a component?
3. If the F1 team is unsure about fitting the sensors, what should they do?
4. If a sensor is checked for accuracy in a test machine, what information will Emma be interested in?

Just checking

1. Why do engineers use standards?

2. What does 'EN' stand for?

3. If a product carries the BSI Kitemark, what does this mean?

4. Which standard gives you the symbols for the components used in pneumatic circuits?

5. If you are studying to become an electronics engineer, can you write down the colour codes for electrical resistors? Probably not. Find them, print them out and make a small reference chart to stick on the wall of the science lab.

6. If you are studying to be a mechanical engineer can you write down the carbon percentages for low-, medium- and high-carbon steels? Again, probably not. Find the figures, and make a small chart for the wall of the science lab.

7. The pages in this book contain text, photographic images, artwork and coloured feature boxes. How do you think this affects the amount of memory needed to store the contents of the book in an electronic filing system?

8. What is the area (measured in m^2) of a sheet of A0 drawing paper?

9. How many times should you fold an A0 size drawing to make it A3 size?

10. After folding a drawing, where should the drawing number present itself?

11. Why are hard copies of drawings printed on standard-sized paper?

edexcel ⠿

Assignment tips

- This unit links with Unit 10: Using computer-aided drawing techniques in engineering. Some of the evidence you have collected can be used to demonstrate how you:

  - produced a circuit drawing using CAD

  - set up an electronic folder for the storage and retrieval of information.

- When you store evidence electronically, make sure you keep backups of your work.

- Drawing standards can be large documents, and sometimes can be difficult to find. Ask your tutor for cut-down printed versions specific to your needs.

- Make sure that you get your tutor to complete observation records or witness statements for any assessment activity that has ephemeral evidence. This could be when you are speaking or doing something in real time that needs to be reported.

- The distinction grading criterion requires that you justify a way of solving a problem with flawed technical information. Justification is about backing up a decision with hard evidence.

3 Mathematics for engineering technicians

To become an engineering technician you need to develop a range of practical skills. You will also need to be able to solve engineering problems and mathematics will help you do this.

Mathematics is an essential tool that helps us to design engineered systems and products and measure their performance. It helps us to assess the strength of engineering materials, to estimate costs and to understand how the different parts of a system or product relate to each other.

In this unit you will learn about the basic rules and methods of arithmetic and algebra and how to apply them in solving problems. Sometimes, when two things are related, it is useful to plot a graph that shows how they affect each other as they change. You will learn how to do this and how to interpret the results.

Very often in engineering we need to calculate areas and volumes. You will learn how to do this and how to apply the skill in estimating material requirements and designing containers. You will also learn how to apply the basic rules and methods of trigonometry to calculate angles and find missing dimensions.

You will need to equip yourself with a scientific electronic calculator and as you progress you will learn how to use it correctly.

Learning outcomes

After completing this unit you should:

1. use arithmetic, algebraic and graphical methods to solve engineering problems

2. use mensuration and trigonometry to solve engineering problems.

Assessment and grading criteria

This table shows you what you must do in order to achieve a **pass**, **merit** or **distinction** and where you can find activities in this book to help you.

To achieve a **pass** grade the evidence must show that you are able to:	To achieve a **merit** grade the evidence must show that, in addition to the pass criteria, that you are able to:	To achieve a **distinction** grade the evidence must show that, in addition to the pass and merit criteria you are able to:
P1 use arithmetic methods to evaluate two engineering problems ensuring answers are reasonable **Assessment activity 3.1** **page 65**	**M1** transpose and evaluate complex formulae **Assessment activity 3.2** **page 72**	**D1** transpose and evaluate combined formulae **Assessment activity 3.2** **page 72**
P2 use algebraic methods to transpose and evaluate simple formulae **Assessment activity 3.2** **page 72**	**M2** identify the data required and determine the area of two compound shapes **Assessment activity 3.3** **page 78**	**D2** carry out chained calculations using an electronic calculator **Assessment activity 3.2** **page 72** **Assessment activity 3.3** **page 78**
P3 plot a graph for linear and non-linear relationships from given data **Assessment activity 3.2** **page 72**	**M3** identify the data required and determine the volume of two compound solid bodies **Assessment activity 3.3** **page 78**	
P4 determine the area of two regular shapes from given data **Assessment activity 3.3** **page 78**	**M4** use trigonometry to solve complex shapes **Assessment activity 3.4** **page 84**	
P5 determine the volume of two regular solid bodies from given data **Assessment activity 3.3** **page 78**		
P6 solve right-angled triangles for angles and lengths of sides using basic Pythagoras' theorem, sine, cosine and tangent functions **Assessment activity 3.4** **page 84**		

How you will be assessed

This unit will be assessed through the responses given to engineering problems and questions that cover the requirements of the assessment criteria and/or by means of investigative assignments. In certain cases your tutor may choose to assess your understanding through oral questions. The assessments will be designed to allow you to show your understanding of the mathematical methods listed in the learning outcomes.

Naomi, 16-year-old apprentice

I have always been interested in practical and science subjects, which is why I opted to do the BTEC Level 2 First in my final two years at school. I liked the way the units were assessed and I achieved merit and distinction grades in most of them. This is equal to four Grade B GCSE passes.

I have never been strong at maths, but it's a subject that you need for a technician career and it gets more interesting as you see how it is used to design and make things. We attended our local further education college for some of the units, where our school didn't have the facilities. I enjoyed the workshop and laboratory sessions and seeing computer aided design and computer aided machining in action.

Our careers teacher arranged for me to apply for an apprenticeship with a company called Fitton Engineering. It's based at our local airport and services light aircraft and helicopters. I passed the aptitude test and interview and now work as part of a small team carrying out servicing and repairs. I am still studying part-time on the BTEC Level 3 National course in Mechanical Engineering. Success on the BTEC Level 2 gave me lots of confidence and I'm now thinking about a technical career in the air force.

Over to you

- What areas of this unit might you find challenging?
- Which section of the unit are you most looking forward to?
- What preparation can you do in readiness for the unit assessment(s)?

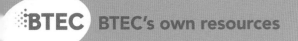

3.1 Be able to use arithmetic, algebraic and graphical methods to solve engineering problems

Start up

What shall we be doing?

Engineers use mathematics in many different ways: for example, carrying out design calculations, estimating the cost of manufacturing a product, analysing test results and checking that they have been paid correctly at the end of the month.

You are now going to work with a colleague to prepare an estimate for repainting the walls of the room that you are currently using.

Start by measuring up the walls so that you can calculate the surface area to be painted. Then work out how many litres of paint are required if a 5-litre can of rich matt will cover approximately 50 m². The can costs £14 - why do you need to know this?

You should include the cost of labour; assume preparation and painting time is 6 minutes per square metre and the pay rate is 2.5 times the UK minimum wage.

Also add in the cost of consumables such as filler, sandpaper and cleaning materials- assume this to be 5% of the paint cost.

So how much will the job cost? Did you allow for one or two coats of paint?

Mathematics is the language of science and technology, but it uses numbers and symbols rather than words. We will start with arithmetic and the way numbers are used in solving engineering problems.

We use algebra to solve problems where we don't know some of the numbers. We call them unknown quantities and use symbols in place of them. Finding them is like solving a puzzle. We use graphs to show how things that are related vary with each other. Graphs plotted from experimental data give us a picture of what has been happening.

3.1.1 Arithmetical methods

We count to the base of 10, probably because we have ten fingers and thumbs. This is called the denary or decimal system. There are other methods of counting. For example, computers use the base of 2 and these are called binary numbers. You will be familiar with the Roman numerals used on clocks and monuments. These were in use for a long time, but are not suited for science and mathematics. The numbers that we use today originated from Arabia and the Middle East and are much easier to read and manipulate than Roman numerals.

Whole numbers, decimal numbers, fractions, percentages

Whole numbers can be positive (+) or negative (−). You can think of money in a savings account as being positive. If you spend it your account will fall to zero and if you carry on spending it will become negative and need to be repaid. We always put the − sign before negative numbers, but we don't always put the + sign before positive numbers. If a number doesn't have a sign before it, you can assume that it is positive.

If a value lies between two whole numbers we can express it in the form of a fraction or a decimal. For instance, half-way between 2 and 3 is $2\frac{1}{2}$ or 2.5. In decimal numbers the interval between whole numbers is split up into 10 and each of these can be sub-divided into 10, depending on the degree of accuracy required. The use of decimal numbers in science and engineering has become almost universal in the UK since it adopted the metric system of measurement. Some common fractions and their equivalent decimal numbers and percentages are shown in Table 3.1.

Table 3.1: Some common fractions and decimals

Fraction	Decimal	Percentage
$\frac{1}{10}$	0.1	10%
$\frac{1}{8}$	0.125	12.5%
$\frac{1}{4}$	0.25	25%
$\frac{1}{2}$	0.5	50%
$\frac{3}{4}$	0.75	75%
$\frac{1}{16}$	0.0625	6.25%
$\frac{1}{100}$	0.01	1%
$\frac{1}{1000}$	0.001	0.1%

When you add and subtract or multiply and divide numbers, there are four basic rules that you must follow.

1. When adding two or more positive numbers, add their **absolute values** together. Sometimes we prefix the answer with a + sign, but usually we leave it out. For example:

$$4 + 3 + 1 = 8 \ \text{(or sometimes +8)}$$

When adding two or more negative numbers, add their absolute values together and prefix the answer with a − sign. For example:

$$-3 + (-2) + (-4) = -9$$

The brackets help us to show clearly that the numbers are negative.

Key term

Absolute value – the numerical value of a number regardless of it's sign

2. To find the sum of a mixture of positive and negative numbers, add together the absolute values of the positive numbers, then add together the absolute values of the negative numbers. Now subtract the smaller total from the larger one and prefix your answer with the sign of the larger number. For example:

$$4 + (-1) + 2 + (-3) = 6 + (-4) = +(6 - 4) = +2$$
$$5 + (-6) + 3 + (-8) = 8 + (-14) = -(14 - 8) = -6$$

3. When you multiply or divide numbers with like signs, the answer will always be positive. For example:

$$5 \times 4 = +20 \text{ and } 16 \div 2 = +8$$
$$(-3) \times (-6) = +18 \text{ and } (-12) \div (-2) = +6$$

Remember: multiplying or dividing two negative numbers always gives a positive answer.

4. When you multiply or divide numbers with unlike signs the answer will always be negative. For example:

$$6 \times (-4) = -24 \text{ and } (-10) \div 2 = -5$$
$$(-7) \times 2 = -14 \text{ and } 18 \div (-2) = -9$$

A percentage (%) is a fraction that has been converted to hundredths. To do this, simply take your fraction or decimal number and multiply it by 100. For example, express $\frac{1}{5}$ as a percentage:

$$\frac{1}{5} \times 100 = 20\%, \text{ which means 20 hundredths}$$

For example, express 0.45 as a percentage:

$$0.45 \times 100 = 45\% \text{ which means 45 hundredths}$$

In Table 3.1, each of the decimal numbers has been multiplied by 100 to express it as a percentage.

Worked example 3.1

Records show that a storage rack contains 120 components. During a working day 15 are taken out, of which 2 are returned unused; then 16 are taken out, of which 3 are returned unused. How many components are left in the rack at the end of the day and what is the percentage reduction in their number?

Finding the final number of components: $120 + (-15) + 2 + (-16) + 3 = 125 + (-31) = 94$ components

Finding the reduction in the number stored: $120 - 94 = 26$ components

Finding the percentage reduction: $\frac{26}{120} \times 100 = 21.7\%$

Worked example 3.2

The scale on an analogue voltmeter reads from zero to 20 volts. When a test voltage of 6 V is applied, the meter reading is 6.15 V. Calculate the error: (a) as a percentage of the true reading; (b) as a percentage of the full-scale deflection.

Finding the error in the reading:

Error = 6.15 − 6 = +0.15 V

(a) Finding the error as a percentage of the true reading:

$$\text{Percentage error} = \frac{+0.15}{6} \times 100 = +2.5\%$$

(b) Finding the error as a percentage of the full-scale deflection:

$$\text{Percentage error} = \frac{+0.15}{20} \times 100 = +0.75\%$$

Ratio and proportion

The **ratio** of two quantities tells you how many times one of them is bigger than the other.

The compression ratio in a petrol or diesel engine gives you a comparison between the initial volume of air in the cylinders and the final volume into which it is compressed. For a typical petrol engine the air might be compressed into $\frac{1}{8}$ of its initial volume and we say that it has a compression ratio of 8 : 1. It is much higher for a diesel engine, which might have a compression ratio of 20 : 1.

$$\text{Compression ratio} = \frac{\text{Initial volume of air}}{\text{Compressed volume of air}}$$

We encounter ratios in many branches of engineering. For example, when a component such as a tie bar is loaded in tension, its length increases by a small amount. This is measured as strain, which is the ratio of the change in length to the original length.

$$\text{Strain} = \frac{\text{Change in length}}{\text{Original length}}$$

In this case we find it more convenient to have the small dimension divided by the large dimension and so strain is always a number less than 1.

When you read an engineering drawing, you will usually see the scale to which it is drawn written in the title block. This, too, is a ratio. It tells you how much bigger the actual component is compared with its size on the drawing.

$$\text{Drawing scale} = \frac{\text{Actual size}}{\text{Drawing size}}$$

Key term

Ratio – compares the sizes of the amounts that a quantity is divided into

Worked example 3.3

The scale of an engineering drawing is given on the title block as 10 mm = 1.5 m. What is the ratio of actual size to drawing size?

Begin by changing the 1.5 m to millimetres: that is, $1.5 \times 1000 = 1500$ mm

Now divide this by 10 mm to give you the ratio:

$$\text{Ratio} = \frac{1500}{10} = 150 \text{ or } 150:1$$

The scale on road maps and atlases is the same kind of ratio. A scale of 500 000:1 is quite common for road maps: this might also be written as 1 cm = 5 km.

Sometimes the ratio between two quantities is given by numbers that are both greater than 1. For instance, you might have to allocate an urgently required batch of components between two customers in the ratio of 3:2 or 5:3 etc. Worked example 3.5 shows how to calculate the **proportion** that each will receive.

Worked example 3.4

The piston of a large internal combustion engine compresses the cylinder contents into a final volume of 0.05 m³ . In doing so, the piston sweeps through a volume of 0.33 m³. Calculate the compression ratio of the engine.

Begin by adding the swept volume to the clearance volume to give the initial volume of the cylinder contents:

Initial volume = 0.33 + 0.05 = 0.38 m³

Now divide this by the final volume to give the compression ratio:

$$\text{Compression ratio} = \frac{\text{Initial volume}}{\text{Final volume}}$$

$$\text{Compression ratio} = \frac{0.38}{0.05} = 7.6 \text{ or } 7.6:1$$

Key term

Proportion – the different amounts that a quantity is divided into.

Worked example 3.5

A batch of 105 components has to be allocated between two customers, A and B, in the ratio 5 : 2. Calculate the proportion that each customer should receive.

Begin by adding together the two numbers in the ratio and dividing it into the number in the batch. This will be a $\frac{1}{7}$th share of the total.

$$\frac{1}{7} \text{ share} = \frac{105}{7} = 15 \text{ components}$$

Customer A will receive 5 of these shares and customer B will receive 2 of them.

Number allocated to customer A = 15 × 5 = 75 components

Number allocated to customer B = 15 × 2 = 30 components

Check your answers by adding these proportions together to make sure that the total is 105 components.

Total = 75 + 30 = 105 components

Therefore, the answers are correct.

1.1.3 Powers and indices

When a number has been multiplied by itself a number of times we say that it has been raised to a **power**. When a number is multiplied by itself we say that it has been **squared** and if it is multiplied by itself again we say that it has been **cubed**.

Taking the number 2 as an example we write these as

$$2^2 = 2 \times 2 = 4$$
$$2^3 = 2 \times 2 \times 2 = 8$$

We can carry on doing this and writing it as

$$2^4 = 2 \times 2 \times 2 \times 2 = 16$$

Here we say that we have raised the number 2 to the power of 4. When we write 2^4, the number 2 is called the **base** and the number 4 is called the **index**. The plural of *index* is *indices*. When the index is 1, this just means the number itself: that is, $2^1 = 2$, $3^1 = 3$ and so on. When the index is zero, it doesn't matter what number the base is; the answer is always 1. That is, $2^0 = 1$, $3^0 = 1$, $4^0 = 1$ and so on.

Key terms

Squared – The square of a number is the number multiplied by itself.

Cubed – The cube of a number is the number multiplied by itself twice.

Power – The power of a number is the number multiplied by itself a number of times.

Base – When we raise a number to a power we call the number the base and the power the **index**.

When we add together or subtract different numbers that have to be raised to a power, we have to raise each one separately and then do the addition or multiplication. For example:

$$2^3 + 3^2 = (2 \times 2 \times 2) + (3 \times 3) = 8 + 9 = 17$$
$$5^2 - 4^3 = (5 \times 5) - (4 \times 4 \times 4) = 25 - 64 = -39$$

The same applies if we are multiplying or dividing different numbers that are raised to a power. For example:

$$2^4 \times 3^2 = (2 \times 2 \times 2 \times 2) \times (3 \times 3) = 16 \times 9 = 144$$
$$4^2 \div 3^2 = (4 \times 4) \div (3 \times 3) = 16 \div 9 = 1.78$$

If, however, we have to multiply powers of the same number (i.e. the same base), we can add the indices together and then raise the number to this power. For example:

$$2^3 \times 2^2 = 2^{(3+2)} = 2^5 = 2 \times 2 \times 2 \times 2 \times 2 = 32$$
$$3^3 \times 3 = 3^{(3+1)} = 3^4 = 3 \times 3 \times 3 \times 3 = 81$$

Note that 3 is taken to be 3^1 in this last example.

When we are dividing powers of the same number, we can subtract the indices before raising the number to this power. For example:

$$2^5 \div 2^2 = 2^{(5-2)} = 2^3 = 2 \times 2 \times 2 = 8$$
$$3^4 \div 3^3 = 3^{(4-3)} = 3^1 = 3$$

It is quite possible to have negative indices. The **reciprocal** of a number always has the index -1. For example:

$$\frac{1}{2} = \frac{1}{2^1} = 2^{-1}, \frac{1}{3} = \frac{1}{3^1} = 3^{-1}, \frac{1}{4} = \frac{1}{4^1} = 4^{-1}, \text{ and so on.}$$

We can do the same with reciprocals of numbers that are raised to a power that is greater than 1. For example:

$$\frac{1}{2^2} = 2^{-2}, \frac{1}{3^4} = 3^{-4}, \frac{1}{4^3} = 4^{-3}, \text{ and so on.}$$

When you want to add, subtract, multiply or divide numbers with negative indices, you need to change them back to reciprocal form first. For example:

$$2^{-3} + 4^{-2} = \frac{1}{2^3} + \frac{1}{4^2} = \frac{1}{2 \times 2 \times 2} + \frac{1}{4 \times 4} = \frac{1}{8} + \frac{1}{16}$$
$$= \frac{2}{16} + \frac{1}{16} = \frac{3}{16} \text{ or } 0.1875$$
$$2^{-2} \times 3^{-1} = \frac{1}{2^2} \times \frac{1}{3^1} = \frac{1}{2 \times 2} \times \frac{1}{3}$$
$$= \frac{1}{4} \times \frac{1}{3} = \frac{1}{12} \text{ or } 0.0833$$

Key term

Reciprocal – The reciprocal of a number is the number divided into 1. That is to say it is made into a fraction with 1 on the top line and the number underneath it.

Roots

The square root of a number is the value that, when multiplied by itself, equals that number. For example, because $2 \times 2 = 4$, the square root of 4 is 2. But it could also be -2, because $(-2) \times (-2) = 4$. The square root is often written as $\sqrt{4}$ or $4^{\frac{1}{2}}$. As you can see, there are two possibilities for the square root of any number. A number such as 4 is called a perfect square, because its square root is a whole number. Other perfect squares are 9, 16, 25, 36, 49 and so on.

You can of course have cube roots and roots to higher powers. Whatever the case, the index is always written as a fraction. For example:

$27^{\frac{1}{3}}$ is the cube root of 27 and is 3

That is, $3 \times 3 \times 3 = 27$

$16^{\frac{1}{4}}$ is the 4th root of 16 and is ± 2

That is, $2 \times 2 \times 2 \times 2 = 16$ and also $(-2) \times (-2) \times (-2) (-2) = 16$

Roots to an uneven power are positive, but roots to an even power can be positive or negative.

Standard form

You can write a number such as 13 579 as 1.3579×10^4. To do this you move the decimal point back to just after the first digit. In this example it has been moved back through 4 places and this is the power of 10 by which the new number must be multiplied to give the same value. It is then said to be written in **standard form**, which is also sometimes called **scientific notation**. The number 1.3579 is called the **mantissa** and the 10^4 is called the **exponent**. For example:

$$254 \times 122.5 = 31\,115 \text{ or } 3.1115 \times 10^4$$

You can do the same with decimal fractions such as 0.01357. This time the decimal point must be moved *forwards* to the right of the first digit, which is 1. It has been moved forward through 2 decimal places and in standard form the number can be written as 1.357×10^{-2}. Note that the exponent is given a minus sign. For example:

$$3.5 \div 1500 = 0.002\,333\,3 \text{ or } 2.3333 \times 10^{-3}$$

You are sometimes asked to give your answer correct to a certain number of decimal places. The above answer is correct to 7 decimal places – that is, 7 places after the decimal point. More often in engineering, though, you are asked to give your answer correct to **3 significant figures**, which is considered to be a sufficient degree of accuracy. This includes the digits before the decimal point. The answer above could then be given as 2.33×10^{-3}. If the fourth significant figure

Key term

3 significant figures – A number given to three significant figures is rounded off to the first three digits. This is not counting the zeros after a decimal point for numbers less than 1.

is 5 or above, the third significant figure can be rounded up.
For example:

$$85 \times 655 = 55\,675 \quad \text{or} \quad 5.57 \times 10^4$$

Engineering notation

The units that we use in engineering, such as millimetres, metres and kilometres, rise in multiples of 1000, which is 10^3. When referring to voltages, you will come across the units mV, V and kV which mean millivolts (mV), volts (V) and kilovolts (kV).

Engineering notation is similar to standard form, except that the mantissa can be between 1 and 999 and the index is always a multiple of 3. This helps us to identify units when solving engineering problems. For example:

$$1500 \text{ m} \times 75 = 112\,500 \text{ m} \quad \text{or} \quad 112.5 \times 10^3 \text{ m} \quad \text{or} \quad 112.5 \text{ km}$$
$$2.47 \text{ V} \div 325 = 0.0076 \text{ V} \quad \text{or} \quad 7.6 \times 10^{-3} \text{ V} \quad \text{or} \quad 7.6 \text{ mV}$$

Key Point

On your scientific calculator you will find a key that is very helpful when entering numbers that are in engineering notation. It is the one that is marked $\boxed{\times 10^n}$ or $\boxed{\text{EXP}}$, which is short for exponent. There is another key marked $\boxed{\text{ENG}}$. We use this for converting our answers into engineering notation.

Worked example 3.6

Use your scientific electronic calculator to evaluate the following expressions. Give your answer in standard form correct to 3 decimal places and in engineering notation correct to 3 significant figures.

(a) $\dfrac{152 \times 180}{0.75} = 36\,480$

In standard form $36\,480 = 3.648 \times 10^4$ correct to 3 decimal places.

To convert your answer to engineering notation, keep the answer on your display and press the $\boxed{\text{ENG}}$ key.
In engineering notation $36\,480 = 36.5 \times 10^3$ correct to 3 significant figures.

(b) $\dfrac{121 \times 10^{-3} \times 350}{2.4} = 17\,645\,833$

To enter 121×10^3, key in $\boxed{1}\boxed{2}\boxed{1}\boxed{\text{EXP}}\boxed{3}$, and then multiply and divide by the other numbers as normal.
In standard form $17\,645\,833 = 1.765 \times 10^7$ correct to 3 decimal places.

To convert your answer to engineering notation, press the $\boxed{\text{ENG}}$ key.
In engineering notation $17\,645\,833 = 17.6 \times 10^6$ correct to 3 significant figures.

(c) $\dfrac{661 \times 10^{-3} \times 25 \times 10^9}{30 \times 10^3} = 550\,833.33$

To enter 661×10^{-3} key in $\boxed{6}\boxed{6}\boxed{1}\boxed{\text{EXP}}\boxed{+/-}\boxed{3}$ and then key in the other numbers to multiply and divide.
In standard form $550\,833.33 = 5.508 \times 10^5$ correct to 3 decimal places.

To convert your answer to engineering notation, press the $\boxed{\text{ENG}}$ key again.
In engineering notation $550\,833.33 = 551 \times 10^3$ correct to 3 significant figures.

Precedence

When you are carrying out a lengthy calculation involving addition, subtraction, multiplication, division and possibly powers and roots, it is essential that you do it in the following order.

1   Work out any terms that are in **B**rackets,

2   Work out any powers or roots (sometimes called '**O**rder').

3   Work out any **D**ivisions and **M**ultiplications (working left to right).

4   Work out any **A**dditions and **S**ubtractions (working left to right).

This is sometimes called the **BODMAS** rule. It is the procedure that your scientific calculator is programmed to follow.

Consider the following calculation:

$$4 - 2 \times 6 + 9 \div 3$$

If you simply work through it in the order in which it is written, the answer comes to 7, but if you do the same on you calculator it comes to −5. This is quite a big difference, but your calculator is correct. There are no brackets, powers or roots, so if you follow the BODMAS rule, the multiplication and division should be done first, giving

$$4 - (2 \times 6) + (9 \div 3) = 4 - 12 + 3 = -5$$

Worked example 3.7

Carry out the following calculations showing the correct order of precedence and check your answers using a scientific calculator.

(a) $(5 + 3) + 6^2 \times (3 - 1) - 12 \div 4$

First, the brackets: $8 + 6^2 \times 2 - 12 \div 4$

Then the powers and roots: $8 + 36 \times 2 - 12 \div 4$

Then the multiplication: $8 + 72 - 12 \div 4$

Then the division: $8 + 72 - 3$

Then the addition: $80 - 3$

Then the subtraction: **77** The answer is 77 and this checks out using the scientific calculator.

(b) $3^3 - (12 + 4)^{\frac{1}{2}} \times (11 - 2) + 12 \div 4$

First the brackets: $3^3 - 16^{\frac{1}{2}} \times 9 + 12 \div 4$

Then the powers and roots: $27 - 4 \times 9 + 12 \div 4$

Then the multiplication: $27 - 36 + 12 \div 4$

Then the division: $27 - 36 + 3$

Then the addition: $27 - 33$ (Be careful, this is $- 36 + 3 = 33$)

Then the subtraction: **− 6** The answer is − 6 and this checks out using the scientific calculator.

Approximations

Sometimes you may need to make a quick estimate of an amount of material that you will need, say, or the time that a particular task will take. You can do this by rounding decimal numbers up or down to the nearest whole number and making a rough mental calculation. It is sometimes called a 'ball-park figure'. It is also useful to do this when you are solving problems, so that if the value you obtain from your calculator turns out to be wildly different, you will know you need to check your calculation.

Worked example 3.8

The operating cost of an automated process is £2.10 per minute. Each component it produces takes 2 min 50 s. What will be the cost per batch of 12?

Finding the approximate time taken to produce 12 components if the time per component is rounded up to 3 min:

Approximate time = 12 × 3 = 36 min

Finding approximate cost if the cost per minute is rounded down to £2:

Approximate cost = 2 × 36 = £72

A more accurate value can be obtained using a scientific calculator, as follows.

Finding time per component in decimal minutes:

$$\text{Time per component} = 2 + \frac{50}{60} = 2.83 \text{ min}$$

Finding time to produce 12 components:

Estimated time = 12 × 2.83 = 33.96 min

Finding estimated cost:

Estimated cost = 2.1 × 33.96 = £71.32

As you can see, there is not much difference between the two, but the ball-park figure can be arrived at quickly without the use of a calculator.

Assessment activity 3.1 P1

1 Water flows up a vertical, tapering pipe. The change in pressure head is given by the formula

$$h = \frac{v_2^2}{2g} - \frac{v_1^2}{2g} + z_2 - z_1$$

where h is the change in pressure head, measured in metres

$v_1 = 1.55\ \text{m s}^{-1}$, the flow velocity at the start of the taper

$v_2 = 2.65\ \text{m s}^{-1}$, the flow velocity at the end of the taper

$z_1 = 5.9\ \text{m}$, the height of the section at the start of the taper

$z_2 = 7.1\ \text{m}$, the height of the section at the end of the taper

$g = 9.81\ \text{m s}^{-2}$, the acceleration due to gravity.

Calculate the change in pressure head correct to three places of decimals.

2 The work input to a mechanical lift is given by the formula

$$W = mg(h_2 - h_1) + \frac{1}{2}m(v_2^2 - v_1^2)$$

where W is the work input, measured in joules

$m = 50\ \text{kg}$, the load raised

$g = 9.81\ \text{m s}^{-2}$, the acceleration due to gravity

$h_1 = 0.5\ \text{m}$, the initial height of the load

$h_2 = 10.2\ \text{m}$, the final height of the load

$v_1 = 1.5\ \text{m s}^{-1}$, the initial velocity of the load

$v_2 = 5.1\ \text{m s}^{-1}$, the final velocity of the load.

Calculate the work input to the lift in engineering notation, correct to three significant figures and express your answer using standard form and scientific notation. P1

Grading tip

Follow the BODMAS rule and show clearly each step in your calculation.

PLTS

Analysing information and evaluating engineering problems will help you to develop your **independent enquirer** skills in handling data.

3.1.2 Algebraic methods

We often express scientific laws and principles as mathematical formulae. They give us the relationship between different physical quantities. Depending on which of the physical quantities we need to solve a problem, we may need to change a formula round to make that quantity the subject. This process is called **transposition**. In some formulae the quantities may be multiplied or divided and there may be roots or powers present. In others the quantities may be added or subtracted; in the more complex formulae there may be a mixture of these operations.

Transposition of formulae

The general rule with formulae in which the quantities are multiplied or divided is that you can transpose them by multiplying or dividing both sides by the quantity that you need to move. For example:

Make v the subject of the formula $s = vt$.

To get v on its own, divide each side by t, so that it can be cancelled on the right-hand side:

$$\frac{s}{t} = \frac{v\cancel{t}}{\cancel{t}}$$

This leaves

$$\frac{s}{t} = v \text{ or } v = \frac{s}{t}$$

For example:

Make R the subject of the formula $I = \frac{V}{R}$.

To get R on its own, begin by multiplying each side by R so that it can be cancelled on the right-hand side.

$$IR = \frac{VR}{R}$$

This leaves

$$IR = V$$

Now divide each side by I, so that it can be cancelled on the left-hand side:

$$\frac{\cancel{I}R}{\cancel{I}} = \frac{V}{I}$$

This leaves

$$R = \frac{V}{I}$$

The general rule with formulae in which the quantities are added or subtracted is that you can transpose them by adding or subtracting from each side the quantity that you need to move. For example:

Make a the subject of the formula $v = u + at$

Begin by subtracting u from each side so that at is left by itself on the right-hand side:

$$v - u = \cancel{u} + at - \cancel{u}$$

This leaves

$$v - u = at$$

Now divide both sides by t so that it can be cancelled on the right-hand side:

$$\frac{v - u}{t} = \frac{a\cancel{t}}{\cancel{t}}$$

This leaves

$$a = \frac{v - u}{t}$$

More complex formulae may contain powers and roots. Where these occur you may have to raise both sides of the formula to a power, or take the root of each side, as a step to isolating the required subject. For example:

Make u the subject of the formula $v^2 = u^2 + 2as$.

Begin by subtracting $2as$ from each side so that u^2 is left by itself on the right-hand side:

$$v^2 - 2as = u^2 + 2as - 2as$$

This leaves

$$v^2 - 2as = u^2$$

Now take the square root of each side:

$$\sqrt{v^2 - 2as} = \sqrt{u^2}$$

This leaves

$$u = \sqrt{v^2 - 2as}$$

Sometimes you may encounter combined formulae that you need to transpose and then evaluate using the data provided. Worked example 3.9 contains such a formula.

Worked example 3.9

Transpose the conservation of energy formula $\frac{1}{2}mv^2 = mgh$ to make the velocity v the subject. Determine the value of v, given that $g = 9.81\ \text{m s}^{-2}$ and $h = 5.1\ \text{m}$.

Begin by cancelling out m, which is on both sides of the formula:

$$\frac{1}{2}\cancel{m}v^2 = \cancel{m}gh$$

This leaves

$$\frac{1}{2}v^2 = gh$$

Now multiply each side by 2 to isolate v^2 on the left-hand side:

$$2 \times \frac{1}{\cancel{2}}v^2 = gh \times 2$$

This leaves

$$v^2 = 2gh$$

Finally, take the square root of each side:

$$\sqrt{v^2} = \sqrt{2gh}$$

This leaves

$$v = \sqrt{2gh}$$

Evaluating when $g = 9.81\ \text{m s}^{-2}$ and $h = 5.1\ \text{m}$:

$$v = \sqrt{2 \times 9.81 \times 5.1}$$

$$= 10.0\ \text{ms}^{-1}$$

Analysing information and evaluating engineering problems will help you to develop your independent enquirer skills in handling data.

3.1.3 Graphical methods

We can picture the way in which two variable quantities are related by plotting their values on a graph. We often do this with data gathered during an experiment . The quantity whose change we can control is called the **independent** variable and the quantity that is changing with it is called the **dependent** variable. We usually plot the independent variable on the horizontal scale (the x-axis) and the dependent variable on the vertical scale (the y-axis). This is not a hard-and-fast rule and if one of the variables is time, we usually plot this on the horizontal scale.

When you are plotting a graph, choose the scale to make the best use of the graph paper, so that the points are not clustered together in one corner. Label the axes of your graph clearly with the names of the variables and their units. Also, give your graph a title that states exactly what has been plotted. Use a pencil to plot the points on your graph carefully and mark each one with a small cross or a dot.

You may find that the points do not quite lie on a straight line or a smooth curve. If any of them seem to be widely displaced, check that you have plotted them correctly. Some scatter may be due to errors in recording equipment or inaccurate measurement when taking readings. When you are satisfied that you have plotted the points correctly, draw a line or curve of best fit that passes through or as close to as many points as possible.

Proportional and linear relationships

When a graph is a straight line passing through the **origin** we say that there is a **proportional** relationship between the variables and the law or equation connecting them is of the form y = mx.

The term m is called the **gradient** or **slope** of the graph. For a relationship such as this it is also called the **constant of proportionality**, which connects the two variables. To find the gradient, draw a large triangle to the graph as in Figure 3.1. Using the axis scales, write on the values **a** and **b** of the vertical and horizontal sides. You can then find the gradient m, using the formula

$$m = \frac{a}{b}$$

When a graph is a straight line that doesn't pass through the origin, we say that there is a linear relationship between the two variables. The law or equation connecting them is now of the form y = mx + c.

Key term

Origin – The point on a graph where the x-axis and y-axis cross.

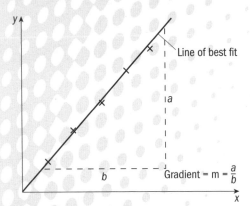

Figure 3.1: This straight-line graph shows a proportional relationship

The gradient *m* is found in the same way and *c* is the value of the intercept on the vertical or *y*-axis. In addition to showing how the two variables are related, you can sometimes find extra information from the area between the graph and the horizontal or *x*-axis. For example, if you are plotting force exerted against distance moved by an object, the area under the graph is the work done. If you are plotting velocity against time, the area under the graph gives the distance travelled.

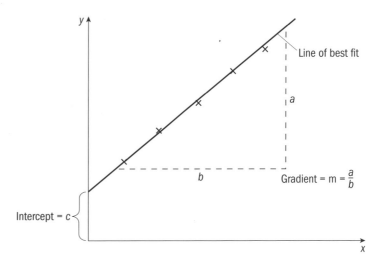

Figure 3.2: This straight-line graph shows a linear relationship

 Worked example 3.10

The following table shows readings of applied load *F* and extension *x* produced during a test on a helical spring.

Load, F (newtons)	0	25	50	75	100	125	150	175	200
Extension, x (mm)	0	0	6	20	34	48	62	76	90

The extension **depends** upon the load applied and the load applied can be **independently** varied as required.

Plot a graph of the two variables and from it determine the stiffness of the spring, measured in N mm⁻¹ and the law connecting the two variables.

The graph is a straight line of the form *W = mx + c*.

The intercept *c* = 40 N. This shows that the spring is close-coiled, i.e that the coils are touching each other, and a load of 40 N is required before the coils start to separate.

The gradient $m = \dfrac{a}{b} = \dfrac{125}{70} =$ 1.79 N mm⁻¹. This is the stiffness of the spring.

The law connecting the two variables is **W = 1.79x + 40**.

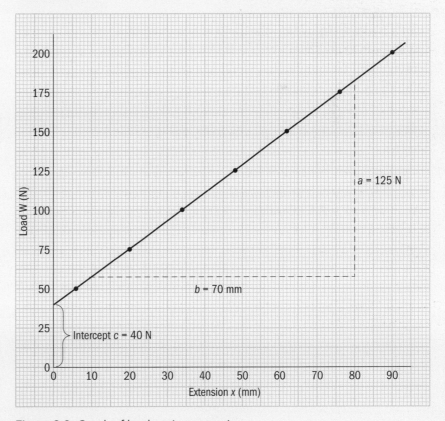

Figure 3.3: Graph of load against extension

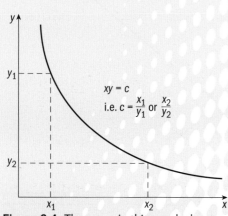

Figure 3.4: The curve in this graph shows an inversely proportional relationship

Non-linear relationships

We sometimes find that quantities are **inversely proportional**. This means that, as one of them increases, the other one decreases by a proportional amount. In such cases the quantities are related by a law of the form $xy = c$, where c is a constant.

If we plot a graph for quantities that vary inversely, the result is a curve, as shown in Figure 3.4.

A gas that obeys Boyle's law behaves in this way. With such a gas an increase in its absolute pressure p causes its volume V to decrease in inverse proportion, provided its temperature remains constant. Pressure and volume are connected by the law $pV = c$, where c is a constant that depends on the particular gas and its mass. The curve that results when quantities such as these are plotted is called a rectangular hyperbola.

Worked example 3.11

The following table shows readings of absolute pressure p of a gas and its volume V, produced during a test carried out at a constant temperature.

Pressure (kPa)	1.0	1.5	2.0	2.5	3.0	3.5	4.0	4.5	5.0
Volume (m³)	2.5	1.67	1.25	1.0	0.83	0.71	0.63	0.56	0.50

Plot a graph of the two variables and from it determine the law that connects them.

From point 1 on the graph, $p_1 = 4.25$ kPa and $V_1 = 0.59$ m³ and so

$$p_1V_1 = 4.25 \times 0.59 = 2.5 \text{ kPa m}^3$$

From point 2 on the graph, $p_2 = 1.25$ kPa and $V_2 = 2.0$ m³ and so

$$p_2V_2 = 1.25 \times 2.0 = 2.5 \text{ kPa m}^3$$

Since $p_1V_1 = p_2V_2 = 2.5$, the law that connects the two variables is **$pV = 2.5$ kPa m³**.

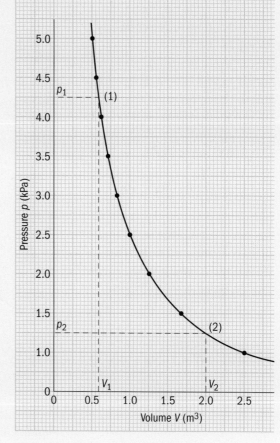

Figure 3.5: Graph of pressure against volume

Another example of a non-linear relationship is provided by quantities that vary according to a **square law**. That is to say, one quantity is proportional to the square of the other. The law is of the form $y = ax^2$, where a is a constant. The curve appears as shown in Figure 3.6.

A typical example of this kind of behaviour is the power consumption in an electric circuit, which is given by the formula $P = I^2R$. Here I is the current and R, the circuit resistance, is the constant. Another example is the kinetic energy of a moving body, given by the formula $E = \frac{1}{2}mv^2$. Here v is the velocity and $\frac{1}{2}m$, which is half the mass of the body, is the constant.

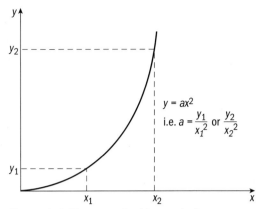

Figure 3.6: The curve in this graph shows a square law relationship

Worked example 3.12

The following table shows readings of the current I flowing in a circuit and the circuit's power consumption P.

Current, I (amps)	0.2	0.4	0.6	0.7	0.8	0.9	1.0	1.2
Power, P (watts)	0.4	1.6	3.6	4.9	6.4	8.1	10.0	14.4

Plot a graph of the two variables and show that they are connected by a square law of the form $P = I^2R$.

If the square law is of the form $P = I^2R$, then $P/I^2 = R$.

From point 1 on the graph, $P_1 = 2$ W and $I_1 = 0.445$ A, and so

$$\frac{P_1}{I_1^2} = \frac{2}{0.445^2} = 10 \ \Omega$$

From point 2 on graph, $P_2 = 12$ W and $I_2 = 1.095$ A, and so

$$\frac{P_2}{I_2^2} = \frac{12}{1.095^2} = 10 \ \Omega$$

Since $P_1/I_1^2 = P_2/I_2^2 = 10 \ \Omega$, the law that connects the two variables is **$P = 10I^2$**.

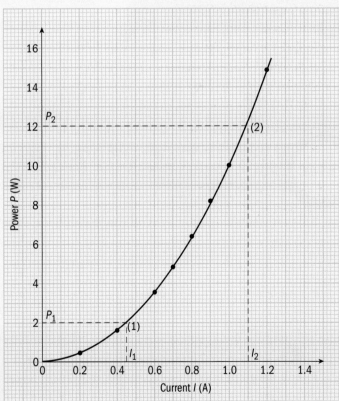

Figure 3.7: Graph of power versus current

When regular increments in one variable produce increasingly large increments in the other, we say that we have **exponential growth**. And when regular decrements in one variable produce ever smaller decrements in the other, we say that we have **exponential decay**. We get examples of these in electrical and mechanical systems and also in nature.

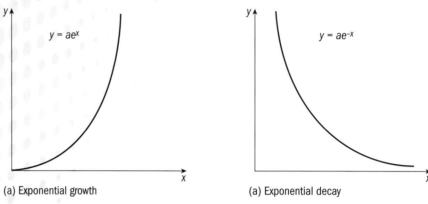

(a) Exponential growth

(a) Exponential decay

Figure 3.8: Graphs of natural growth and decay

For instance, the human population of the world has risen exponentially over the centuries, apart from a few setbacks due to wars, plagues and famine. We have seen an exponential decay in the population of some animals, such as the Asian tiger and the giant panda, which are threatened with extinction. When the figures are plotted on graphs we call them **natural growth** and **decay curves**. They appear as shown in Figure 3.8.

Natural growth and decay curves have characteristic equations that relate the two variables. They are $y = ae^x$ and $y = ae^{-x}$. The constant a can have any positive or negative value, but the constant e is special and always has the value (to three decimal places) 2.718. Just as $\pi = 3.142$ is a constant peculiar to circular measurement, so $e = 2.718$ is a constant peculiar to natural growth and decay graphs.

PLTS

Analysing information and plotting graphs will help you to develop your **independent enquirer** skills in handling data.

BTEC **Assessment activity 3.2**   (P2) (P3) (M1) (D1) (D2)

1. In an electrical circuit the voltage, current and resistance are related by the formula

   V = IR

   If V = 5 volts and R = 47 kΩ calculate the value of the current. (P2)

2. The displacement of an object in a given time is as shown in the formula below.

   $$S = \frac{1}{2}(u + v)t$$

   Transpose the equation to make t the subject and then find its value if $s = 10$ m, $u = 2$ m s^{-1} and $v = 6.5$ m s^{-1} (P2)

3. The force F exerted by a jet of water as it strikes a stationary flat plate at right angles is given by the formula

   $$F = \frac{\rho \pi d^2 v^2}{4}$$

   where the density of water $\rho = 1000$ kg m^{-3}, the diameter of the jet $d = 0.05$ m and $\pi = 3.142$. Determine the velocity v in m s^{-1} of the jet that will deliver a force of 450 N. (M1)

4 The instantaneous voltage of an alternating supply is given by the following equation: $V = V_0 \sin 2\pi ft$

Find a value for t when $V = 80$ volts if $V_0 = 110$ volts and $f = 50$ Hz

Express your answer in standard form **M1**

5 When a mass free falls in a gravitational field its motion is controlled by the following relationship:

$$\frac{1}{2}mv^2 = mgh$$

If $g = 9.81 ms^{-2}$, $m = 4$ kg and $h = 10$ m calculate its velocity v **D1** **D2**

6 The charge relationship for a capacitor is given by the following expression:

$$\frac{1}{2}QV = \frac{1}{2}CV^2$$

Find V if $Q = 750 \times 10^{-6}$ coulombs and $C = 6 \times 10^{-6}$ farads **D1** **D2**

7 When an object is pushed across a flat, dry surface the following formulae relate to the amount of energy needed to do this:

$E = F_f \times s$ and $F_f = \mu mg$

If $E = 1.57$ kJ, $m = 80$ kg, $s = 5$ m and $g = 9.81$ ms^{-2} what is the value of μ? **D1** **D2**

8 The following table of load and effort readings was obtained from a test on a lifting device.

Effort, E (N)	20.5	25.0	29.5	35.0	40.5	45	49.5	55.0	60.0
Load, W (N)	200	400	600	800	1000	1200	1400	1600	1800

Plot a graph of effort (y-axis) against load (x-axis). Use it to determine the effort needed to overcome friction in the mechanism and the law that connects effort and load. **P3**

9 A test to determine the maximum safe working load that a beam can carry as its span is increased returned the following results.

Span, L (m)	0.5	0.75	1.0	1.25	1.5	1.75	2.0	2.25	2.5
Load, W (kN)	2.25	1.50	1.13	0.90	0.75	0.64	0.56	0.50	0.45

It is thought that the safe working load varies inversely with the span. Plot a graph of load (y-axis) against span (x-axis). Use it to show that there is an inverse relationship between the variables and determine the law that connects them. **P3**

Grading tips

- Re-read the section on powers and roots before you attempt the transposition in task 3. Show clearly each step in your transposition and make full use of your electronic calculator in your evaluation. Carry out the evaluation twice to confirm your answer. **M1**

- In tasks 5 and 6 you will need to handle powers and roots in this transposition and evaluation. Show each step clearly in your transposition and recheck it. Make full use of the power and root keys on your electronic calculator. Carry out the evaluation twice to confirm your answer. **D1** **D2**

- In tasks 5, 6, and 7 you must demonstrate to your tutor/assessor how you used a calculator to obtain your answers.

- In tasks 8 and 9 choose suitable scales that make best use of your graph paper. Draw a line and a curve of best fit. Show clearly how you have arrived at the laws connecting the variables. **P3**

Pay attention to presentation.

3.2 Be able to use mensuration and trigonometry to solve engineering problems

Mensuration is the branch of mathematics dealing with measurement – in particular, the measurement of areas and volumes. In engineering we might be required to calculate the area of sheet metal required for a product, or the volume of a containing vessel such as a storage tank or a boiler. The shapes that we have to deal with may be simple or complex. However, it is often possible to break a mensuration problem down into a number of simple shapes or volumes. These can then be calculated and totalled up to solve the problem.

Trigonometry is the study of the relationship between the sides and angles of triangles. Don't be put off by the word. A triangle only has three sides and three angles, so it can't be *too* complicated. We shall put trigonometry to good use in finding missing dimensions that we may need to calculate areas and volumes.

3.2.1 Area

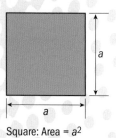
Did you know?

An average football pitch is approximatley 1 hectare (2 to 2.5 acres).

Areas are generally measured in square metres (m²), except for very small areas, for which we may use square centimetres (cm²) or square millimetres (mm²) and very large areas of land, which are measured in hectares (ha). A hectare is 10000 or 10⁴ square metres.

Areas of regular shapes

By regular shapes we mean squares, rectangles, triangles and circles. The formulae for finding their areas are shown in Figure 3.9.

Square: Area = a^2

Rectangle: Area = ab

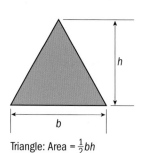

Triangle: Area = $\frac{1}{2}bh$

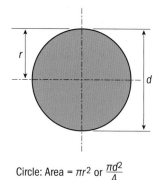

Circle: Area = πr^2 or $\frac{\pi d^2}{4}$

Figure 3.9: Areas of regular shapes

Areas of compound shapes

More complex shapes can often be split up into these basic regular shapes so that you can find their area. Figure 3.10 shows how you can find the area of a rhombus, a parallelogram and a trapezium by rearranging them into squares and rectangles.

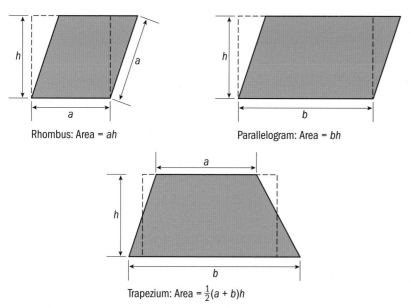

Rhombus: Area = *ah*

Parallelogram: Area = *bh*

Trapezium: Area = $\frac{1}{2}(a + b)h$

Figure 3.10: Using regular shapes to find the area of more complex shapes

As you can see from Figure 3.10, a rhombus is like a distorted square and a parallelogram is like a distorted rectangle. If we remove the triangular area (shown dotted) from the right-hand side and attach to the left-hand side, this converts both shapes into rectangles. We can then find their areas by multiplying their base by their perpendicular height.

A trapezium is a four-sided figure with two sides that are parallel and two that are not. In Figure 3.10, the vertical dotted lines intersect the sloping sides at their mid-points. If we remove the triangular corner sections and reattach them to the upper side, we get a rectangle whose length is the average of the two parallel sides. We can then find the area by multiplying the average length by the distance between the parallel sides.

You will regularly encounter other complex shapes in sheet metal work. The panels that are used to make washing machines, ovens and cookers contain holes, triangular off-cuts and curves. Worked example 3.13 shows you how to find the area of such a panel.

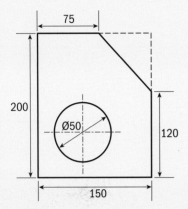

Worked example 3.13

Calculate the surface area of the sheet metal panel shown in Figure 3.11. All the dimensions are in millimetres.

Figure 3.11

The panel will begin life as a rectangular sheet. The triangular corner section will then be removed and the 50 mm diameter hole cut out.

Finding the area of the initial rectangular sheet:

Area of rectangle = Base × Height = 200 × 150

Area of rectangle = 30 × 10³ mm²

Finding dimensions of triangular corner off-cut:

Base = 150 − 75 = 75 mm

Perpendicular height = 200 − 120 = 80 mm

Finding area of triangular off-cut:

Area of triangular off-cut = $\frac{1}{2}$ Base × Perpendicular height = $\frac{1}{2}$ × 75 × 80

Area of triangular off-cut = 3 × 10³ mm²

Finding area of circular hole:

Area of hole = $\frac{\pi d^2}{4} = \frac{\pi \times 50^2}{4}$

Area of hole = 1.96 × 10³ mm²

Finding final area of panel:

Area of panel = Area of rectangle − Area of triangle − Area of hole

Area of panel = (30 × 10³) − (3 × 10³) − (1.96 × 10³) = (30 − 3 − 1.96) × 10³

Area of panel = 25.04 × 10³ mm²

3.2.2 Volume

Large volumes are generally measured in cubic metres (m³). For smaller volumes we may use litres and for still smaller volumes we may use cubic centimetres (cm³). It will be useful to remember that 1 m³ = 1000 litres (10³ l).

1 litre = 1000 cm³ (10³ cm³)

1 m³ = 1 000 000 cm³ (10⁶ cm³)

Volumes of regular solids

By regular solids we mean right-rectangular prisms, cylinders, cones and spheres. The formulae for finding their volumes are shown in Figure 3.12.

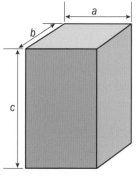 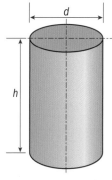 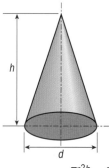 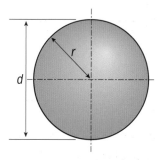

Rectangular prism: Volume = abc Cylinder: Volume = $\pi r^2 h$ or $\dfrac{\pi d^2 h}{4}$ Cone: Volume = $\dfrac{\pi r^2 h}{3}$ or $\dfrac{\pi d^2 h}{12}$ Sphere: Volume = $\dfrac{4\pi r^3}{3}$ or $\dfrac{\pi d^3}{6}$

Figure 3.12: How to find the volumes of these regular solid bodies

Worked example 3.14

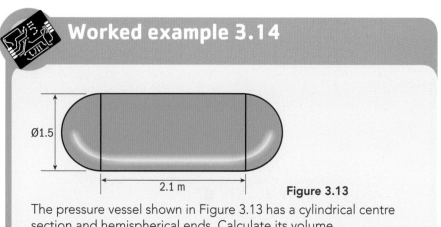

Ø1.5

2.1 m   **Figure 3.13**

The pressure vessel shown in Figure 3.13 has a cylindrical centre section and hemispherical ends. Calculate its volume.

Finding volume of centre cylindrical section:

Volume of centre cylindrical section = $\dfrac{\pi d^2 h}{4} = \dfrac{\pi \times 1.5^2 \times 2.1}{4}$

Volume of centre cylindrical section = 3.71 m³

Finding volume of hemispherical ends, i.e. equal to that of a sphere:

Volume of hemispherical ends = $\dfrac{\pi d^3}{6} = \dfrac{\pi \times 1.5^3}{6}$

Volume of hemispherical ends = 1.77 m³

Finding total volume of pressure vessel:

Total volume = Volume of centre section + Volume of ends

Total volume = 3.71 + 1.77 = 5.48 m³

1. Calculate the areas of the hollowed rectangle and ring shown in Figure 3.14. **P4**

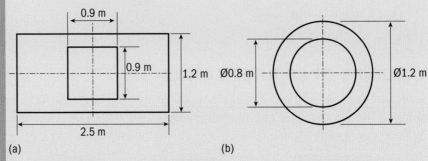

(a)   (b)

Figure 3.14

2. Calculate the areas in square metres of the sheet metal panels shown in Figure 3.15. Make a list of the data you used to do this. **M2**

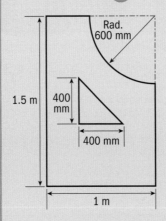 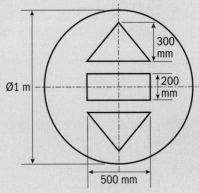

Panel is symmetrical about horizontal and vertical centre lines

(a)   (b)

Figure 3.15

3. Calculate the volumes of the hollow cylinder and hollow sphere shown in Figure 3.16. **P5**

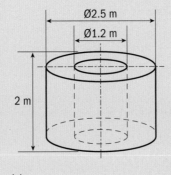 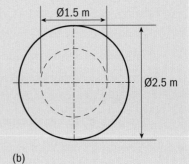

(a)   (b)

Figure 3.16

4. Calculate the volumes of the spherical pressure vessel and the conical hopper shown in Figure 3.17. Make a list of the data you used to do this. **M3** **D2**

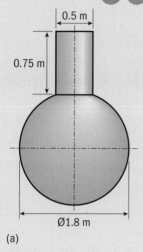
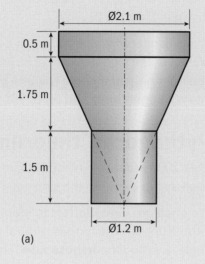

(a)　　　　　　　　　　　(a)

Figure 3.17

Grading tip

Work tidily, show all your working and make full use of your scientific electronic calculator. Check your calculations and check with your tutor the degree of accuracy required in your answers.

PLTS

Analysing information and calculating volumes will help you to develop your **independent enquirer** skills in handling data.

3.2.3 Trigomometry

We use trigonometry a lot in engineering design to calculate dimensions and angles. Building and land surveyors also use it to calculate distances and heights. As we have said, it is the study of the relationship between the sides and angles of triangles. Some triangles have special names, as shown in Figure 3.18.

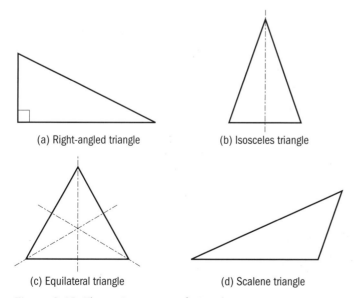

(a) Right-angled triangle　　　　(b) Isosceles triangle

(c) Equilateral triangle　　　　(d) Scalene triangle

Figure 3.18: The various types of triangle

In an equilateral triangle the sides are all the same length and all the angles are equal. An isosceles triangle has two sides the same length and two of its angles are equal. In a scalene triangle the sides are all unequal. We shall be dealing mostly with right-angled triangles, as shown in Figure 3.19.

The circumference of a circle can be divided up into 360 degrees (360°). Whatever the kind of triangle, its internal angles always add up to 180°. The right-angled triangle has one angle that is 90° and so the other two internal angles must add up to 90°.

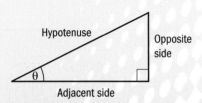

Figure 3.19: A right-angled triangle

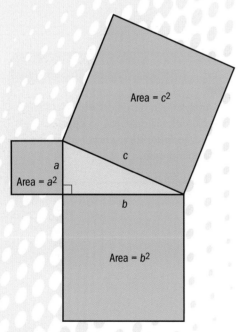

Figure 3.20: Pythagoras' theorem

Pythagoras' theorem

Over 2000 years ago in ancient Greece, Pythagoras discovered the relationship between the sides of a right-angled triangle.

We often use Greek letters to identify angles. Here we shall use the letter θ (theta). The longest side of the triangle, opposite the right angle, is called the **hypotenuse**. The side **opposite** the angle θ in Figure 3.19 is called the opposite side and the other side, next to the angle θ, is called the **adjacent** side.

Pythagoras discovered that if you make each side of a right-angled triangle the side of a square, as in Figure 3.20, the area of the square on the hypotenuse is equal to the sum of the areas of the squares on the other two sides.

This means that

$$c^2 = a^2 + b^2$$

or

$$c = \sqrt{a^2 + b^2}$$

Pythagoras' theorem comes in very useful for solving right-angled triangles when we know the length of two sides and need to find the remaining one.

An easy way to construct a right-angled triangle on paper is to use the 3, 4, 5 method. Note that $3^2 + 4^2 = 5^2$, so the sides of your triangle will be in these relative lengths.

Start by drawing the hypotenuse 5 units long, then set your compasses to 4 units and strike an arc from one end of the hypotenuse. Now set your compasses to 3 units and strike another arc from the other end of he hypotenuse. Where the two arcs intersect gives you the point of the right angle and enables you to draw in the two remaining sides.

Tangent of an angle

An acute angle is one that is less than 90°. The angles at A and B in Figure 3.21 are acute angles. Angle C is 90° and the sides opposite the angles A, B and C are lettered a, b and c.

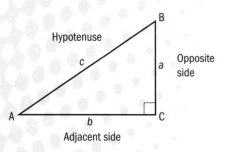

Figure 3.21: Right-angled triangle notation

The ratio of side a, opposite angle A, to side *b*, adjacent to angle A, is known as the **Tangent** of the angle. It is usually written as tan A for short.

$$\tan A = \frac{\text{Opposite side}}{\text{Adjacent side}}$$

$$\tan A = \frac{a}{b}$$

You can find the tangent of any angle on your electronic calculator by entering the number of degrees and pressing the tan key. If you know the angle and either of the sides, you can find the other one. Alternatively, if you know both the opposite and adjacent sides, you can calculate the tangent of the angle θ and obtain its value in degrees by pressing the shift tan keys.

Worked example 3.15

In the triangle ABC shown in Figure 3.22, calculate the angle at A, the angle at B and the length of the hypotenuse c.

$$\text{Tan A} = \frac{a}{b} = \frac{1.5}{2.5}$$

$$\text{Tan A} = 0.6$$

Finding value of angle i.e. by pressing the shift tan keys.

A = 30.96°

Finding angle at B.

$$A + B + C = 180°$$

$$B = 180 - C - A = 180 - 90 - 30.96$$

B = 59.04°

Finding value of hypotenuse c, using Pythagoras' theorem.

$$c = \sqrt{a^2 + b^2} = \sqrt{1.5^2 + 2.5^2}$$

c = 2.9 m

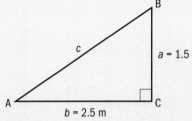

Figure 3.22: Finding tangent of angle at A

2.3.3 Sine of an angle

The ratio of the side *a*, opposite the angle A, to side *c*, the hypotenuse, is known as the **Sine** of the angle. It is usually written as Sin A for short.

$$\text{Sin A} = \frac{\text{Opposite side}}{\text{Hypotenuse}}$$

$$\textbf{Sin A} = \frac{a}{c}$$

You can find the Sine of any angle on your electronic calculator by entering the number of degrees and pressing the sin key. If you know the angle and either of the sides *a* or *c*, you can find the other one. Alternatively, if you know both the opposite side and the hypotenuse you can calculate the Sine of the angle A, and obtain its value in degrees by pressing the shift sin keys.

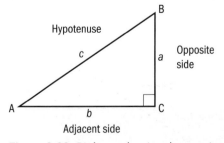

Figure 3.23. Right-angle triangle notation

Worked example 3.16

In the triangle ABC shown in Figure 3.24, calculate the length of the sides b and c and the angle at B.

Finding side c.

$$\text{Sin } A = \frac{a}{c}$$

$$c = \frac{a}{\text{Sin } A} = \frac{2.1}{\text{Sin } 30°}$$

$$c = \frac{2.1}{0.5} = \textbf{4.2 m}$$

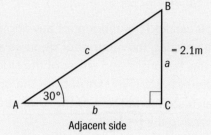

Figure 3.24

Finding side b.

$$\text{Tan } A = \frac{a}{b}$$

$$b = \frac{a}{\text{Tan } A} = \frac{2.1}{\text{Tan } 30°}$$

$$b = \frac{2.1}{0.577} = \textbf{3.64 m}$$

Checking the answers using Pythagoras' theorem

$$c = \sqrt{a^2 + b^2} = \sqrt{2.1^2 + 3.64^2}$$

$$c = \textbf{4.2 m} \quad \text{i.e the answers check out}$$

Finding angle at B.

$$A + B + C = 180°$$

$$B = 180° - A - C = 180 - 30 - 90$$

$$\textbf{B = 60°}$$

A triangle such as this is known as a 60°–30° triangle. You should note that with such a triangle the hypotenuse is always twice the length of the shortest side.

2.3.4 Cosine of an angle

The ratio of the side b, adjacent to the angle A, to side c, the hypotenuse, is known as the **Cosine** of the angle. It is usually written as Cos A for short.

$$\text{Cos } A = \frac{\text{Adjacent side}}{\text{Hypotenuse}}$$

$$\textbf{Cos } A = \frac{b}{c}$$

You can find the Cosine of any angle on your electronic calculator by entering the number of degrees and pressing the 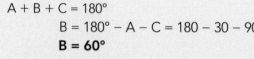 key. If you know the angle A and either of the sides b or c, you can find the other one. Alternatively, if you know both the adjacent side and the hypotenuse you can calculate the Cosine of the angle A, and obtain its value in degrees by pressing the [shift] [cos] keys.

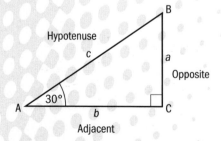

Figure 3.25

Worked example 3.17

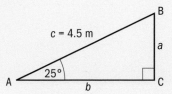

Figure 3.26

In the triangle ABC shown in Figure 3.26, calculate the length of the sides b and c and the angle at B.

Finding side b.

$$Cos\ A = \frac{b}{c}$$

$$b = c\ Cos\ A = 4.5\ Cos\ 25°$$

$$\mathbf{b = 4.5 \quad 0.906 = 4.08\ m}$$

Finding side a.

$$Sin\ A = \frac{a}{c}$$

$$a = c\ Sin\ A = 4.5\ Sin\ 25°$$

$$\mathbf{a = 4.5 \times 0.423 = 1.90\ m}$$

Checking the answers using Pythagoras' theorem

$$c = \sqrt{a^2 + b^2} = \sqrt{1.90^2 + 4.08^2}$$

$$\mathbf{c = 4.50\ m} \quad \text{i.e. the answers check out}$$

Finding angle at B.

$$A + B + C = 180°$$

$$B = 180° - A - C = 180 - 25 - 90$$

$$\mathbf{B = 65°}$$

2.3.5 Trig. relationship

As you now know, the tangent, sine and cosine of the angle at A are calculated as follows.

$$Tan\ A = \frac{a}{b} \qquad\qquad Sin\ A = \frac{a}{c} \qquad\qquad Cos\ A = \frac{b}{c}$$

Dividing Sin A by Cos A gives,

$$\frac{Sin\ A}{Cos\ A} = \frac{a}{c} \div \frac{b}{c} = \frac{a}{\cancel{c}} \times \frac{\cancel{c}}{b}$$

$$\frac{Sin\ A}{Cos\ A} = \frac{a}{b}$$

But this is the same as Tan A, and so we have the relationship,

$$\mathbf{Tan\ A = \frac{Sin\ A}{Cos\ A}}$$

You won't make much practical use of this relationship immediately but you may find that it comes in useful in future studies.

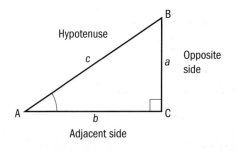

Figure 3.27: Right-angled triangle notation

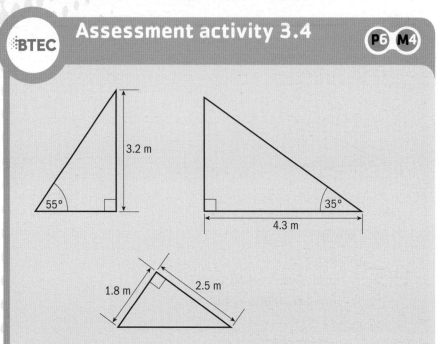

Figure 3.28

1. Calculate the angles and side dimensions for each of the triangles shown in Figure 3.28. Your calculations must include use where appropriate of Pythagoras' theorem and sine, cosine and tangent functions. **P**6

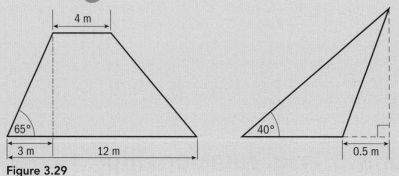

Figure 3.29

2. Calculate the height, sloping side lengths and angles of the trapezoid shaped roof truss.

   Calculate the angles and side dimensions of the triangle.

Grading tips

Show all your working and make full use of your scientific electronic calculator. Wherever possible check your answers using an alternative method of calculation. Check with your tutor the degree of accuracy required in your answers.

WorkSpace Andrew Griffin

Air-conditioning technician

I work for a company who design, manufacture and install air-conditioning systems to hospitals, public buildings and factories. I started work as an apprentice and studied part-time at my local Further Education College. I am now a fully qualified air-conditioning technician and work as part of a team engaged in the design and costing of full air-conditioning systems. My responsibilities include identifying customers' air-conditioning requirements, choosing the equipment, designing the ducting and estimating the cost. They also include scheduling the installation of the system.

A typical day for me might begin by meeting an architect to discuss the air-conditioning requirements of the various workspaces, corridors and reception areas of a new building. There are strict regulations that we have to adhere to so that a building is fit for its purpose and in which people feel comfortable at work or when visiting.

Depending on the size of the contract, my colleagues and I might spend several days calculating air-flow rates and selecting the most suitable equipment for the task. A typical system contains a lot of pipes and rectangular ducts that deliver fresh air and carry away the return air. These are all made from sheet metal and plastic materials. Some are bought in as finished items but there are always some special sections which have to be designed and made in our workshops.

All of the material requirements, equipment and installation costs have to be estimated and totalled up as accurately as possible. There are always some problems that arise during the installation of a system but working from past experience, we try to make provision for these. The task involves the calculation of areas, volumes, times, percentages and costs. When this is done, our firm can put forward a realistic tender for the contract. If the tender is accepted we can then draw up an installation schedule and place orders for materials and equipment with our suppliers.

Think about it!

1. What background knowledge and skills have you covered in this chapter that would be used by Andrew when he designs an air-conditioning system? Write a list and discuss with your peers.

2. What further skill do you think you would need to develop so that you could work successfully as part of a team and meet with customers? Write a list and discuss in small groups.

Just checking

1. Explain the BODMAS rule

2. Explain the difference between stating an answer correct to three places of decimals and three significant figures

3. What do you use the [exp]EXP and [eng] keys on your calculator for?

4. Describe two kinds of non-linear relationship

5. How do you calculate the area of a parallelogram?

6. How do you calculate the volume of a sphere?

7. What is Pythagoras' theorem?

8. What is the relationship between the Tangent, Sine and Cosine of an angle?

edexcel

Assignment tips

- Keep a tidy file of notes and worked examples that you can refer to when revising for an assessment

- Recognise the areas where you are least confident and don't hesitate to seek help from your tutors

- Don't leave things until the last minute. To become proficient in mathematics you have to keep practising the different techniques. This is especially true in the case of transposition of formulae. It is good practice to word process your assignment work. It will develop your IT skills and make it easier for your tutor to assess your work and give you feedback.

Credit value: 5

4 Applied electrical and mechanical science for engineering

Mechanical engineering has a long history. The Romans developed machines for lifting huge stone blocks for building and also military machines for hurling large stones at their enemies. In the Middle Ages advances were made in the design of wind and water mills and clockmakers invented some of the complex mechanisms that we still use today. Thanks to Sir Isaac Newton and other mathematicians and scientists it was at this time that the laws and principles that underpin engineering were discovered. This made further advances possible and we still use those principles today in the design of mechanical systems.

Electrical engineering has a shorter history. It was only around 200 years ago that Michael Faraday and others discovered the laws and principles that underpin the design and operation of electric motors, generators and electrical circuits.

A lot of the things that we take for granted today contain both electrical and mechanical systems. The range includes domestic appliances, cars and aircraft that depend on an efficient combination of the two disciplines.

In this unit you will be introduced to some of the basic scientific principles that underpin electrical and mechanical engineering and how they are applied to solve engineering problems. They are rather like building blocks: once they are in place, they will enable you to move on to a higher level of study.

Learning outcomes

After completion of this unit you should be able to:

1. define and apply concepts and principles relating to electrical science

2. define and apply concepts and principles relating to mechanical science.

Assessment and grading criteria

This table shows you what you must do in order to achieve a **pass**, **merit** or **distinction** and where you can find activities in this book to help you.

To achieve a **pass** grade the evidence must show that you are able to:	To achieve a **merit** grade the evidence must show that, in addition to the pass criteria, you are able to:	To achieve a **distinction** grade the evidence must show that, in addition to the pass and merit criteria, you are able to:
P1 define parameters of direct current electricity and magnetic fields **Assessment activity 4.1 page 99**	**M1** determine the force on a current carrying conductor situated in a magnetic field from given data **Assessment activity 4.2 page 108**	**D1** explain the construction, function and use of an electro-magnetic coil **Assessment activity 4.2 page 108**
P2 determine total resistance, potential difference and current in series and parallel dc circuits from given data **Assessment activity 4.2 page 108**	**M2** describe the conditions required for the static equilibrium of a body. **Assessment activity 4.4 page 126**	**D2** determine the work done and the power dissipated in moving a body of given mass along a horizontal surface at a uniform velocity, given the value of the coefficient of kinetic friction between the contact surfaces. **Assessment activity 4.4 page 126**
P3 define parameters of static and dynamic mechanical systems **Assessment activity 4.3 page 112**		
P4 determine the resultant and equilibrant of a system of concurrent coplanar forces from given data **Assessment activity 4.4 page 126**		
P5 determine the uniform acceleration/retardation of a body from given data **Assessment activity 4.4 page 126**		
P6 determine the pressure at depth in a fluid from given data. **Assessment activity 4.4 page 126**		

How you will be assessed

This unit will be assessed by your tutor by means of investigative assignments and/or through the responses given to engineering problems and questions that cover the requirements of the assessment criteria.

Your assessment will most likely be in the form of a number of written assignments in which you will need to give the definitions of electrical and mechanical system parameters and provide the solution to a number of related engineering problems. In certain cases your tutor may choose to assess your understanding through oral questions.

Phillip, 16-year-old engineering apprentice

Success on the BTEC First Diploma course gave me a real sense of achievement. I studied for my First Diploma during my last two years at school. Part of the course was delivered at the local college of further education, which has superb engineering facilities and with which my school has strong links. I passed the course with Merit/Distinction grades, which is equivalent to four GCSE passes at Grade B.

I particularly enjoyed the workshop sessions and the computer aided design. I found that I had to work hard on the course, especially with the mathematics and science units, but with some extra help from the tutors I managed to achieve Merit grades in these units. The college tutors offered guidance in some areas and suggested that I apply for an apprentice position with a local company called Terolux, which specialises in supplying and servicing industrial kitchen and food processing equipment. My application was successful and I was taken on as an apprentice. My main tasks are to assist in the production of sheet metal components and fabrications.

The BTEC First Diploma course certainly helped me to develop the personal skills that I need when working as part of a team. I enjoy my job very much and am continuing to study part-time at college on the BTEC National Certificate course in mechanical engineering.

Over to you

- What area of this unit do you think you will find challenging?
- Which section of the unit do you think you will find the most interesting?
- How can you best prepare for unit assessment?

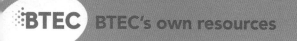

4.1 Be able to define and apply concepts and principles relating to electrical science

Start up

Could we manage without it?

Think of the different ways that you have used electricity today and how different your life would be without it. How do you think you could use it more efficiently?

Write down four different ways in which electricity is generated and consider which you think is the most environmentally friendly.

Discuss your findings in small groups. Identify where you might be able to make savings and which methods of generation you would recommend to meet future increases in demand.

4.1.1 Definitions of parameters of direct current

A parameter of a system is a quantity:

- that is related to the way the system operates
- that helps to describe the state of the system
- that may change and can be measured.

Some of the parameters that relate to direct current electricity are as follows.

Electrical charge

To understand what makes an electric current, you first need to know about electrical charge. The atoms that make up all substances are composed of a nucleus that carries a positive charge and orbiting electrons that carry a negative charge. The basic unit of electrical charge is that carried by a single electron, but this is extremely small and the SI unit of charge is the **coulomb (C)**, where

1 coulomb = 6.25×10^{18} electron charges

Electrical current (*I*)

When a conductor such as a lamp is attached across the terminals of a battery, a current of electrons is made to flow through it. The flow is continuous in one direction and that is why we call it a direct current. The SI unit of electric current is the **ampere (A)**, where

1 ampere = 1 coulomb of charge passing in 1 second

We use the symbol I for current. An ammeter connected in a circuit records the current flowing.

Electromotive force (E) and potential difference (V)

Electromotive force (emf) is the force exerted by a device such as a battery or dynamo that causes motion of the electrical charges. The SI unit of electromotive force is the **volt (V)**. We often refer to the potential difference (V) between the ends of a conductor. This is the change in voltage due to the energy being used to pass current against the resistance of the conductor. Electromotive force, as applied to batteries, dynamos and generators, is the voltage available at their output terminals before any current is drawn from them. When a current flows, however, the terminal voltage is seen to fall a little and what remains is the potential difference across the connected circuit. A voltmeter connected across a circuit, or part of a circuit, records the potential difference that is present.

Electrical resistance (R)

The flow of electrons in a conductor is impeded by the stationary atoms around which they must pass. We call this electrical resistance. The SI unit of electrical resistance is the **ohm (Ω)**.

A conductor has a resistance of 1 ohm when a potential difference of 1 volt causes a current of 1 ampere to flow.

Electrical power

When an electric current flows along a conductor we can say that some work has been done, or that some energy has been transferred. **Electrical power** is that rate of energy transfer. Its SI unit is the watt (W), which is the rate at which energy is transferred when a potential difference of 1 volt causes a current of 1 ampere to flow. It is calculated by multiplying the potential difference V volts by the current I amperes:

Electrical power = $V I$ (watts)   (1)

The purpose of an electrical system is to deliver this energy in a useful form. Sometimes it is used to power a motor and in other instances it is used to produce light, heat or sound.

Did you know?

Although a lamp lights up immediately, the drift of electrons that form the current is very slow. It is said that, for a typical electric torch, no one electron manages to make the complete trip around the circuit during the life of the battery.

Key terms

Electromotive force (emf) – the force which causes motion of the electrical charges.

Potential difference – the change in voltage due to energy being used to pass current through some resistance.

Electrical power – the rate at which energy is transferred. It is measures in watts (W). Heat, light and work are forms of energy into which electrical energy is often transferred.

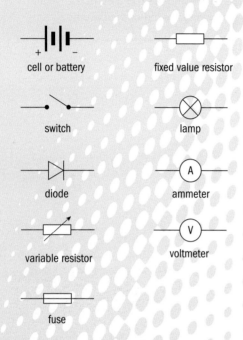

cell or battery   fixed value resistor

switch   lamp

diode   ammeter

variable resistor   voltmeter

fuse

Figure 4.1: Some common electric circuit symbols

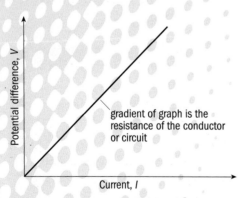

gradient of graph is the resistance of the conductor or circuit

Figure 4.2: Graph of potential difference against current

4.1.2 Direct current electric circuits

An electric circuit consists of several components linked together and connected to a source of electromotive force such as a battery.

Circuit symbols

In order to understand electric circuits you need to be able to recognise the symbols used for the circuit components. Some of the more common ones are shown in Figure 4.1.

Ohm's law

Ohm's law gives us the relationship between potential difference and current. We state it as follows:

> The current flowing in a conductor is proportional to the potential difference between its ends.

It assumes that physical conditions such as temperature do not change and that if they do, the flow of current will not be affected. When Ohm's law is obeyed, a graph of potential difference against current for a conductor or circuit will be a straight line that starts from the origin, as shown in Figure 4.2.

As Ohm's law states, the graph shows that current is proportional to potential difference. That is:

Potential difference = Constant × Current

The constant is the resistance R of the conductor and so we can write

Potential difference = Resistance × Current

or

$$V = IR \qquad (2)$$

This simple formula is the foundation of all electrical calculations. It can be transposed to make either the current I or the resistance R the subject. That is:

$$I = \frac{V}{R} \qquad (3)$$

and

$$R = \frac{V}{I} \qquad (4)$$

As you can see from Figure 4.2, you can also find the resistance of a conductor or circuit from the gradient of the graph of potential difference against current.

You may find it easy to remember these formulae by drawing the triangle shown in Figure 4.3.

We can use these Ohm's law formulae to give us alternative ways of calculating electrical power. From equation 1:

Electrical power = VI

But $V = IR$, so substituting for V gives

Electrical power = $IR \times I$

That is,

Electrical power = I^2R (5)

Also, $I = V/R$. Substituting gives

Electrical power = $V \times \dfrac{V}{R}$

That is,

Electrical power = $\dfrac{V^2}{R}$ (6)

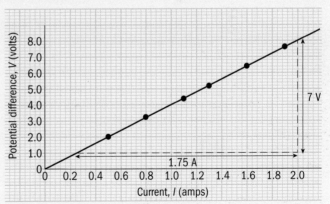

Figure 4.3: Ohm's law triangle

Worked example 4.1

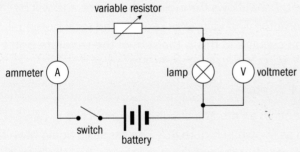

Figure 4.4

In the circuit in Figure 4.4 the variable resistor is used to vary the current flowing through the lamp, making it brighter or dimmer. The voltmeter is used to record the potential difference across the lamp. The following readings of current and potential difference are obtained.

Current (A)	0.5	0.8	1.1	1.3	1.6	1.9
Potential difference (V)	2.0	3.2	4.4	5.2	6.4	7.6

Plot a graph of potential difference against current and use it to find the resistance of the lamp.

When plotted, the graph is a straight line through the origin, as shown in Figure 4.5. This shows that Ohm's law is obeyed i.e. that the current flowing through the lamp is proportional to the potential difference across it.

Figure 4.5: Graph of potential difference against current

The gradient of the graph, which gives the resistance of the lamp, is found by drawing the triangle shown and dividing the vertical side (7 V) by the horizontal side (1.75 A).

Finding the resistance of the lamp:

Resistance of lamp = Gradient of graph

$R = \dfrac{7}{1.75} = 4\,\Omega$

Worked example 4.2

An ammeter placed in an electric circuit shows that a current of 0.25 A is flowing and a voltmeter placed across the input terminals to the circuit records a potential difference of 9.5 V. Calculate: (a) the resistance of the circuit; (b) the power dissipated.

(a) Finding the resistance of the circuit:

$$R = \frac{V}{I} = \frac{9.5}{0.25}$$

$$R = 38 \ \Omega$$

(b) Finding the power dissipated (remember that this means the energy transferred per second):

Power = VI = 9.5 × 0.25

Power = 2.38 W

Resistance in series and parallel circuit networks

When the components in an electric circuit are connected in series, they are connected nose to tail, as shown in Figure 4.6. Each resistor has a resistance (R_1, R_2, R_3 etc.) and the total resistance R of the circuit is the sum of the separate resistances:

$$R = R_1 + R_2 + R_3 + \dots \text{ etc.} \tag{7}$$

It is the same current, I, that flows through each resistor and will be recorded on the ammeter. The voltmeter records the total potential difference across the circuit and this is the sum of the separate potential difference values V_1, V_2, V_3 etc. across each resistance:

$$V = V_1 + V_2 + V_3 + \dots \text{ etc.} \tag{8}$$

When the components in an electric circuit are connected in parallel, they are connected as shown in Figure 4.7. The resistors again have the resistances R_1, R_2, R_3 etc., but the resultant resistance R of the circuit is found from the sum of their reciprocals:

$$\frac{1}{R} = \frac{1}{R_1} + \frac{1}{R_2} + \frac{1}{R_3} + \dots \tag{9}$$

Now it is the same potential difference V across each resistor, as recorded on the voltmeter, but the current through each one is different. The total current I is the sum of the separate currents I_1, I_2, I_3 etc.:

$$I = I_1 + I_2 + I_3 + \dots \text{ etc.} \tag{10}$$

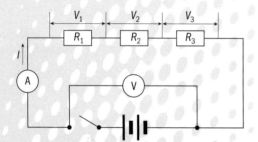

Figure 4.6: Resistors connected in series

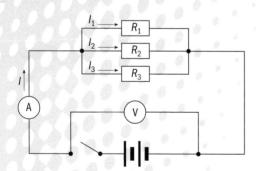

Figure 4.7: Resistors connected in parallel

Did you know?

When resistances are connected in parallel, the resulting resistance of the network is always less than the smallest value of resistance. Why do you think that this is so?

Worked example 4.3

A circuit contains three resistors in series, which are connected to a 12 V supply as shown in Figure 4.8. Calculate: (a) the resistance R of the circuit; (b) the current I flowing through the resistors; (c) the potential difference across each of the resistors; (d) the power dissipated in the circuit.

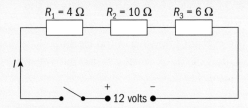

Figure 4.8

(a) Finding the total resistance of the circuit:

$R = R_1 + R_2 + R_3 = 4 + 10 + 6$

$R = 20\ \Omega$

(b) Finding the current flowing in the circuit:

$I = \dfrac{V}{R} = \dfrac{12}{20}$

$I = 0.6\ A$

Now that we know the current flowing through the resistors, we can apply Ohm's law formula for potential difference to each one in turn.

(c) Finding the potential difference V_1 across R_1:

$V_1 = IR_1 = 0.6 \times 4$

$V_1 = 2.4\ V$

Finding the potential difference V_2 across R_2:

$V_2 = IR_2 = 0.6 \times 10$

$V_2 = 6.0\ V$

Finding the potential difference V_3 across R_3:

$V_3 = IR_3 = 0.6 \times 6$

$V_3 = 3.6\ V$

Finally, do a check to see that these all add up to the supply voltage:

$V = V_1 + V_2 + V_3 = 2.4 + 6.0 + 3.6$

$V = 12\ V$ i.e. the answers are correct

(d) Finding the power dissipated:

Power $= I^2R = 0.6^2 \times 20$

Power $= 7.2\ W$

Worked example 4.4

A circuit contains two resistors connected in parallel to a 9 V supply as shown in Figure 4.9. Calculate: (a) the resistance R of the circuit; (b) the current I flowing in the circuit; (c) the current flowing through each of the resistances; (d) the power dissipated in the circuit.

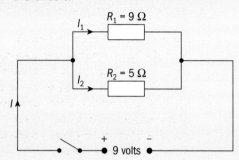

Figure 4.9

(a) Finding the resulting resistance R of the circuit:

$$\frac{1}{R} = \frac{1}{R_1} + \frac{1}{R_2} = \frac{1}{9} + \frac{1}{5}$$

This is where your electronic calculator comes in handy. Use your ⃞1/x key to add these reciprocals together as follows:

⃞9 ⃞1/x ⃞+ ⃞5 ⃞1/x ⃞=

The answer that should appear on your display is

$$\frac{1}{R} = 0.311$$

You can now press your ⃞1/x key once again to invert this number, giving you the value of R. It should appear on your display as

$R = 3.21\ \Omega$

As you can see, this is less than the value of the lower of the two resistances.

(b) Finding the current I flowing in the circuit:

$$I = \frac{V}{R} = \frac{9}{3.21}$$

$I = 2.8\ A$

(c) Finding the current I_1 flowing through resistance R_1:

$$I_1 = \frac{V}{R_1} = \frac{9}{9}$$

$I_1 = 1.0\ A$

Finding the current I_2 flowing through resistance R_2:

$$I_2 = \frac{V}{R_2} = \frac{9}{5}$$

$I_2 = 1.8\ A$

Finally, do a check to see that these all add up to the total current flowing:

$$I = I_1 + I_2 = 1.0 + 1.8$$

$I = 2.8\ A$ i.e. the answers are correct

(d) Finding the power dissipated in the circuit:

$$\text{Electrical power} = \frac{V^2}{R} = \frac{9^2}{3.21}$$

$\text{Electrical power} = 25.2\ W$

4.1.3 Definitions of parameters of magnetic fields

There is a very close relationship between magnetic fields and electric current. Working in the 1780s in Bologna, Italy, the Italian scientist Alessandro Volta produced the first battery, enabling the flow of direct current to be studied. The unit of potential difference is named after him. It was observed that a magnetic field is set up around a conductor carrying a current and scientists began to wonder whether the reverse might be possible. That is, could a magnet be used to produce an electric current. It was not until the 1830s that Michael Faraday, working in England, solved the problem and invented the world's first dynamo.

Magnetic fields

You will be familiar with fridge magnets and probably also with horseshoe magnets and bar magnets. A magnet seems to affect the space around it. We say that it produces a force field that we can't see, but we know exists, because the magnet will attract iron and steel objects. Magnets will also attract or repel each other.

When a bar magnet is suspended by a cord at its centre it will swing round and align itself with the earth's magnetic field. The end that points north is designated a north pole because it seeks north and the other end is designated a south pole. Sometimes they are labelled N and S. This is what happens with the magnetised needle of a compass.

A magnetic field is set up wherever charged particles are in motion. We have already said that a direct electric current is the flow of electrons along a conductor. This is the reason why a magnetic field is set up around a current-carrying conductor. When the conductor is in the form of a coil wound around an iron core the magnetic field can become very strong. This is the principle behind the construction of electromagnets.

In every kind of material, electrons are in motion around the nucleus of the atoms and each electron produces a small magnetic field. In most cases, however, they orbit in pairs in such a way as to cancel out each other's magnetism. But in some materials, such as iron, nickel and cobalt, there are unpaired electrons where the magnetic fields are not cancelled out. When these electron orbits are aligned in the same direction, a strong magnetic field can be produced. This is what happens when iron or steel become magnetised to make a horseshoe magnet or a bar magnet.

Did you know?

It is more correct to call the north pole of a bar magnet a north-seeking pole because it will point towards the earth's north pole if freely suspended. If unlike poles attract, is the north pole of the magnet really a south pole? You might like to think about this and discuss it with your class.

Magnetic flux (φ)

We say that a magnetic field is made up of lines of force. When you sprinkle iron filings onto a sheet of paper under which there is a bar or horseshoe magnet, they seem to arrange themselves along the lines of force. We call these imaginary lines of force **magnetic flux** (φ). The lines are not really present, but it is convenient to think of a magnetic field in this way. The SI unit of magnetic flux is the **weber** (**Wb**), which you can think of as belonging to a single line of force.

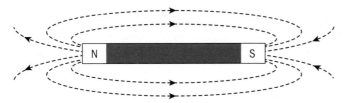

Figure 4.10: The magnetic field around a bar magnet

The magnetic flux is assumed to flow out of the north pole of a magnet and into the south pole, as shown in Figure 4.10. In the case of a straight current-carrying conductor, the flux encircles it as shown in Figure 4.11a. Its direction is said to follow a right-hand screw rule: that is, when the current is flowing away from the observer, the flux is in a clockwise direction.

The way that we indicate the direction of current flow and direction of the lines of force is shown in Figure 11b. Imagine the cross (+) as being the head of a wood screw. To screw it into a piece of wood you would turn it clockwise and that is the direction of the lines of force when the current is flowing away from you. On the far right diagram, imagine the point of the screw coming towards you. It will be turning anticlockwise as you see it and that is the direction of the lines of force when the current is flowing towards you.

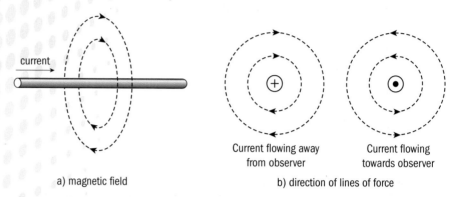

a) magnetic field

Current flowing away from observer

Current flowing towards observer

b) direction of lines of force

Figure 4.11: Magnetic field around a current-carrying conductor

A current-carrying coil produces a magnetic field similar to that of a bar magnet. The coil is known as a **solenoid**. The north and south pole ends are determined by the direction of the windings and the direction of the current, as shown in Figure 4.12. When the coil is wound around an iron core and the current is switched off, the core is found to have retained some of its magnetism, becoming a bar magnet. Soft iron tends to lose its magnetism fairly quickly, but hard steel will keep it for long periods. This is how bar magnets are made.

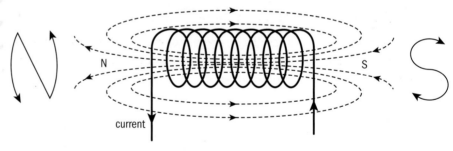

Figure 4.12: Magnetic field around a current-carrying coil

Flux density (*B*)

In a strong magnetic field the lines of force are packed closely together and we measure the intensity of the field by its **flux density** (B). This is the number of lines of force (Wb) passing at right angles through an area of 1 square metre. The SI unit of flux density is the **tesla** (T), which is a flux of 1 weber per square metre.

Did you know?

An iron or steel magnet will lose its magnetism when heated to 770°C, which is known as its Curie temperature.

Did you know?

The direction of the lines of force around a current-carrying conductor is such as to obey a **right-hand screw rule**, i.e. clockwise when the current is flowing away from the observer.

Assessment activity 4.1

Read carefully through your notes that define the parameters of direct current and magnetic fields:

- electric charge
- electric current
- electromotive force
- electrical resistance
- electric power
- magnetic flux
- magnetic flux density

Then, without referring to your notes, test yourself by writing down the definition and unit of measurement for each of these parameters. Check your answers and give yourself 1 point for a correct definition and 1 point for the correct unit of measurement. **P1**

Scores:

12–14	Excellent
9–12	Good
7–9	Fair, but more revision needed
Less than 7	a better understanding needed before assessment by your tutor.

4.1.4 Magnetic circuits

Just as electric current flows around electric circuit, we say that magnetic flux can flow around a magnetic circuit. Figure 4.13 shows the two types of circuit.

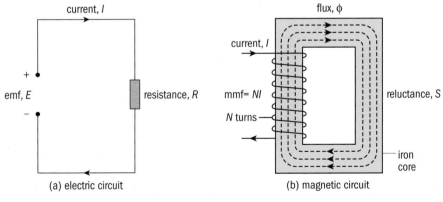

(a) electric circuit   (b) magnetic circuit

Figure 4.13: Electric and magnetic circuits

In place of electromotive force (emf) we have **magnetomotive force (mmf)**, which is the current flowing through the coil multiplied by the number of turns. In place of resistance we have **reluctance (S)**, which depends on the material on which the coil is wound and in place of electric current we have the **magnetic flux** (φ). Soft iron has a very low reluctance compared with most other metals and non-metals. As a result the flux is much stronger with an iron core. You may go on in later work to carry out calculations on magnetic circuits that involve these parameters.

Electromagnetic induction

Michael Faraday discovered that when a conductor is moved through a magnetic field in such a way that it cuts through the flux, a potential difference is set up between its ends. The process is called **electromagnetic induction** and in later studies you will be shown how to calculate the induced emf.

Faraday went on to design a dynamo known as Faraday's disc, which is shown in Figure 4.14. The flux between the poles of a horseshoe magnet is cut by a rotating copper disc. Brushes on the perimeter and the axle of the disc pick up the induced emf. All modern dynamos and electrical generators operate on this principle, but the disc is replaced by a system of rotating coils.

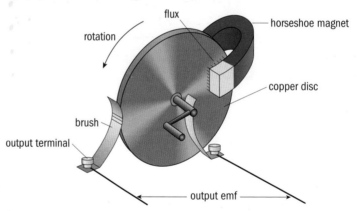

Figure 4.14: Faraday's disc

So far we have considered only direct currents, but our domestic supply is alternating current (ac). That is to say that the current is constantly changing direction and this occurs 50 times per second. We say the mains frequency is 50 hertz (Hz). The domestic supply voltage is 230 V, but the power stations send out electricity at a much higher voltage. This is progressively stepped down in transformer stations before it reaches us.

Transformers

Electromagnetic induction also occurs if we have a stationary conductor and a magnetic field that moves so as to be cut by it. When an alternating current flows in a coil it sets up a magnetic field in which the flux is constantly changing. If another coil is placed close to it, the changing flux passes through it as well. We say that the **flux linkage** between the coils is changing, which results in an emf being induced in the second coil. The process is called **mutual induction** and is the principle on which transformers operate.

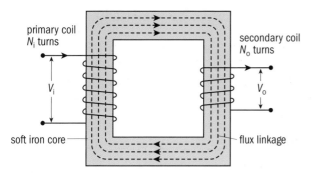

Figure 4.15: The basic principle of a transformer

A transformer consists of an iron core on which two coils are wound, as in Figure 4.15. They are called the **primary coil**, which is connected to the ac supply or input voltage V_i and the **secondary coil**, which delivers the required output voltage V_o. Remember, transformers can only work with alternating current. The symbol used for a transformer on circuit diagrams is shown in Figure 4.16.

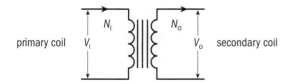

Figure 4.16: The circuit diagram symbol for a transformer

The size of the output voltage depends on the number of turns on the secondary coil, N_o. If this is less than the number on the primary coil, N_i, it will be smaller than the input voltage. The opposite occurs if the secondary coil has more turns than the input coil: in this case the transformer will step up the input voltage. The ratio of the two voltages is equal to the ratio of the numbers of turns, giving us the formula

$$\frac{V_i}{V_o} = \frac{N_i}{N_o} \tag{11}$$

 Did you know?

The earth's magnetic field is called its magnetosphere.

Worked example 4.5

A transformer with 1500 turns on its primary coil is connected to a 230 V ac supply. (a) Calculate the number of turns on its secondary coil if the output voltage is 9 V. (b) Calculate the output voltage if the secondary coil is replaced by one containing 457 turns.

(a) Finding the number of turns:

$$\frac{V_i}{V_o} = \frac{N_i}{N_o}$$

Transposing to make N_o the subject of the formula gives

$$N_o = \frac{V_o N_i}{V_i} = \frac{9 \times 1500}{230}$$

$N_o = 59$ *turns*

(b) Finding the output voltage:

$$\frac{V_i}{V_o} = \frac{N_i}{N_o}$$

Transposing to make V_o the subject of the formula gives

$$V_o = \frac{V_i N_o}{N_i} = \frac{230 \times 457}{1500}$$

$V_o = 70$ V

Force on a conductor in a magnetic field

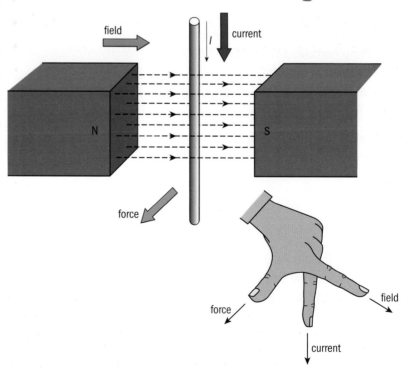

Figure 4.17: Force on a conductor in a magnetic field

When a current-carrying conductor passes through a magnetic field as shown in Figure 4.17, it experiences a force that tries to push it sideways. The reason is that the circular magnetic field produced around the conductor reacts against the field through which it passes. This is shown in Figure 4.18.

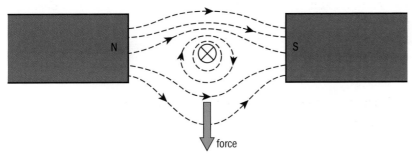

Figure 4.18: The reason for the force on the conductor

The directions of the magnetic field, the current and the force exerted are given by **Fleming's left hand rule**, which is shown in Figure 4.17. When you hold your left hand in the position shown, the direction of the flux is given by the first finger. The direction of the current *I* is given by the middle finger and the direction of the thrust is given by the thumb.

The force, or thrust, acting on the conductor is proportional to three things:

- the flux density *B* of the magnetic field, which is measured in tesla (T)
- the current *I* flowing in the conductor, which is measured in amperes (A)
- the length *L* of the conductor that is in the field, which is measured in metres.

Multiplying these together will give you the value of the force F on the conductor, measured in newtons (N):

$$F = BIL \tag{12}$$

Worked example 4.6

(a) Calculate the force acting on a conductor that carries a current of 12 A and is situated in a magnetic field of flux density 0.3 T at right angles to the flux. The length of the conductor that lies in the field is 1.5 m. (b) Calculate the current needed to increase the force on the conductor to 10 N.

(a) Finding the force on the conductor:

$F = BIL = 0.3 \times 12 \times 1.5$

$F = 5.4$ N

(b) Finding the current required:

$F = BIL$

$I = \dfrac{F}{BL} = \dfrac{10}{0.3 \times 1.5}$

$I = 22.2$ A

DC motors

We have described Faraday's disc, which is the simplest form of dynamo. Now that we know about the force on a conductor, we can investigate the simplest form of direct current (dc) motor.

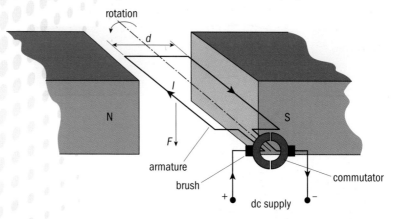

Figure 4.19: The simplest form of dc motor

The rotating part of an electric motor is called the **armature**. In this case it is a rectangular coil, which is fed with direct current through the brushes and **commutator**. This simple commutator has two segments. When the motor is in the position shown in Figure 4.19, each long side of the coil experiences a force F that causes the armature to rotate. The forces exert a turning moment that causes the coil to rotate. We call this the output **torque**, which is the force multiplied by the turning radius r.

The torque is at its maximum when the coil is in the position shown. As the coil rotates, the force stays the same, pulling upwards and downwards, but the torque becomes less. When the coil is in the vertical position the force F is still upwards and downwards, but there is no turning effect and the torque falls to zero.

Because the force F acts on each side of the coil, the maximum torque will be

> Maximum torque = $2Fr$

But $2r = d$, the width of the coil, so that,

> Maximum torque = Fd (13)

Now, from equation (12), $F = BIL$, so that

> Maximum torque = $BILd$

The torque can be increased by having more than one turn of wire on the coil. If it has n turns, the maximum torque will be

> Maximum torque = $nBILd$ (14)

As the coil passes through the vertical position, the two halves of the slip-ring exchange brushes. This ensures that the current is always in the direction that maintains the rotation and the torque builds up to a maximum again.

Did you know?

The force on a current-carrying conductor is proportional to the flux density of the magnetic field, the effective length of the conductor and the current flowing through it.

Moving-coil meters

Analogue voltmeters and ammeters, in which a pointer moves over a scale to give a reading, are generally of the moving-coil type. The coil, which is shown in Figure 4.20, is wound around a soft iron core. The principle of operation is very similar to the dc motor. The difference is that the coil, to which the pointer is attached, is allowed to swing only through an angle of around 90°.

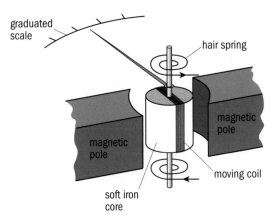

Figure 4.20: Moving-coil meter

The two hairsprings have two purposes. They control the swinging of the pointer and are also used to supply current to the coil. The difference between a voltmeter and an ammeter is the resistance that each offers to the flow of current. A voltmeter has a high resistance whereas an ammeter has a low resistance. You will learn the reasons for this when you continue your studies at a higher level.

Worked example 4.7

The armature coil of a simple dc motor has 250 turns of wire and the magnetic field between the pole pieces has a flux density of 0.3 T. The effective length of the coil sides is 120 mm and the width of the coil is 80 mm. Calculate the force acting on each side of the coil when a direct current of 3.5 A flows through it. Also calculate the maximum torque acting on the coil.

Finding the force acting on each side of the coil:

$F = nBIL$ (where n = 250 turns)

$F = 250 \times 0.3 \times 3.5 \times 0.12$

$F = 31.5\ N$

Finding the maximum torque acting on the coil:

Maximum torque = Fd (where d = 0.08 m)

Maximum torque = 31.5×0.08

Maximum torque = $2.52\ Nm$

Solenoid applications

The solenoid, which is basically a coil of wire wound around an iron core, is put to use in a great many engineering applications. The magnetic field it produces is used to open and close electrical switches and also pneumatic and hydraulic valves. We have already described electric motors and moving-coil meters. Moving coils are also to be found in microphones and loudspeakers.

Relays and contactors

A **relay** is a magnetically operated switch. It enables a small electric current in one circuit to switch a much larger electric current on or off in another circuit. A relay consists of a coil of wire wound on a soft iron core, a yoke and a hinged armature, as shown in Figure 4.21a.

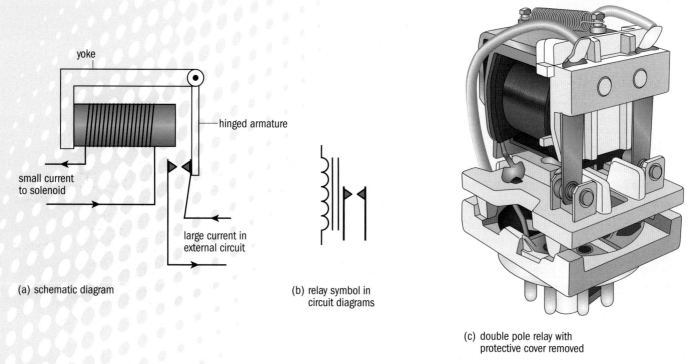

(a) schematic diagram

(b) relay symbol in circuit diagrams

(c) double pole relay with protective cover removed

Figure 4.21: Relays

When you turn the ignition key to start a car engine you activate a switch similar to that shown schematically in Figure 4.21a. It is very often referred to as the 'solenoid', but it is in fact a relay. It allows a very small current from the battery to switch on a much larger current to the starter motor.

Heavy-duty relays called **contactors** are used to switch high-voltage power supply systems on and off. They operate on the same principle, but are much larger and of more robust construction.

Solenoid valves

Just as relays are used to control the flow of electric current, solenoid valves are used to control the flow in pneumatic and hydraulic circuits. In these valves the iron core of the solenoid is a cylinder that contains a spring-loaded piston, as shown in Figure 4.22.

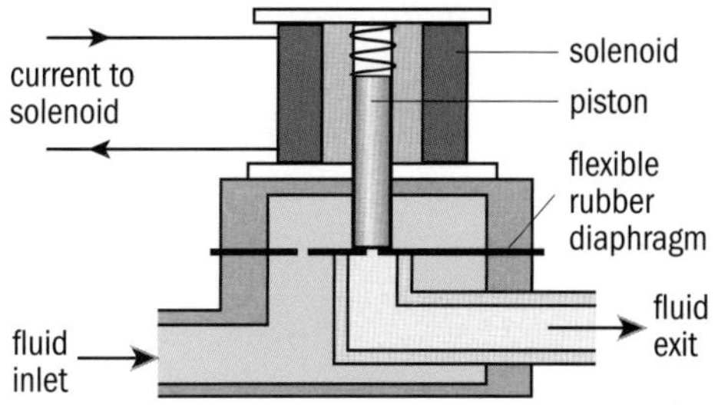

When the solenoid is energised, the magnetic field pushes the piston up the cylinder to compress the spring. As this occurs, the piston releases a diaphragm, allowing the fluid to flow. The diaphragm contains two small holes. The one on the left in the diagram equalises the pressure above and below it when the valve is in the closed position. As the piston rises, the hole in the centre allows a little fluid to flow from above it, so that the pressure is greater underneath the diaphragm. This causes the diaphragm to flex upwards, allowing the fluid to flow to the exit. The spring returns the piston to its original position when the current to the solenoid is switched off and the flow ceases.

This is just one type of solenoid valve. With other types used in pneumatic and hydraulic actuation circuits the piston is arranged to cover and uncover ports in the wall of a cylinder. This directs fluid in different directions to operate the different actuators. Automated packaging and machining systems use this type of solenoid valve.

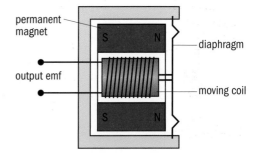

Figure 4.22: Solenoid valve

Microphones and loudspeakers

Loudspeakers and some types of microphone contain a solenoid. They are called moving-coil devices. You will remember from Faraday's disc that when a conductor cuts through lines of force, an emf is induced in it. If the conductor is in the form of a coil, the same thing occurs. This is what happens in a moving-coil microphone.

As can be seen from Figure 4.23, the coil is surrounded by a permanent magnet and attached to a sensitive diaphragm.

Sound waves from speech or music cause the diaphragm to vibrate and this vibration is transmitted to the moving coil. As the coil vibrates it cuts the magnetic flux and an emf is generated. This 'copies' the movement of the diaphragm and is passed to an external circuit for processing.

A moving-coil loudspeaker is of similar construction but operates in the reverse way. The moving coil is again surrounded by a permanent magnet and is wound around a conical diaphragm, as shown in Figure 4.24.

The input signal is in the form of a fluctuating current that corresponds to the speech or music that is being received. The fluctuating current gives rise to a corresponding fluctuating force on the coil that causes it to vibrate from side to side. The diaphragm picks up the vibrations and converts them into sound waves.

Figure 4.23: Moving-coil microphone

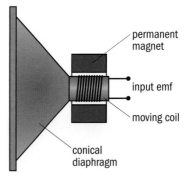

Figure 4.24: Moving-coil loudspeaker

Assessment activity 4.2

BTEC

1. For each of the circuit networks in Figure 4.25 determine:

   - the current flowing from the supply
   - the potential difference across each resistance
   - the current flowing through each resistance.

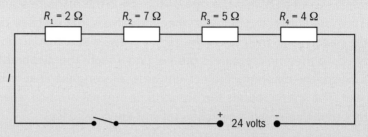

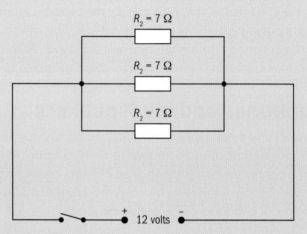

Figure 4.25

2. A conductor is situated in a magnetic field at right angles to the flux. The cross-sectional area of the magnetic field is 0.5 m², with a uniform flux of 2.4 Wb. The length of the conductor affected by the field is 210 mm. The conductor is part of a circuit whose resistance is 9 Ω, and is connected to a 120 V supply. Calculate the force acting on the conductor.

3. Explain the construction, function and use of an electro-magnetic coil. Illustrate your answer with suitable diagrams. **D1**

PLTS

Analysing information and solving problems relating to electrical circuit networks will help you to develop your **independent enquirer** skills in handling data.

Functional skills

Selecting and applying formulae to solve electrical circuit network problems and checking your answers will help you to develop your **mathematical skills**.

Grading tip

Work tidily, with neat and accurately labelled diagrams. Set out your calculations in a logical order and show all your working. Make full use of your electronic calculator and always repeat a calculation to confirm your answer. If there is an alternative formula for the parameter you have calculated, use it as a check.

4.2 Be able to define and apply concepts and principles relating to mechanical science

4.2.1 Definitions of parameters of static and dynamic systems

Scalar and vector quantities

A vector quantity is a measure of something that has both magnitude (size) and a direction. An example of this is the force acting on an object. Its magnitude is measured in newtons, but just as important is the direction in which it acts. A vector quantity can be represented by a line drawn to a suitable scale on a vector diagram. The length of the line represents the magnitude of the quantity and the way that it points represents its direction.

A scalar quantity is one that has magnitude only: that is, it just has a size. An example of this is the mass of an object. It is measured in kilograms and doesn't have any direction.

Mass

Mass is a measure of the amount of matter contained in a body. The SI (Système International) unit of mass is the kilogram (abbreviation kg). It is defined as the mass of a platinum-iridium cylinder measuring 39.17 mm in both height and diameter that resides in the International Bureau of Weights and Measures at Sèvres on the outskirts of Paris.

Density

The density of a substance is its mass per cubic metre. Its symbol is the Greek letter ρ ('rho'). In the case of water, 1 litre has a mass of 1 kg and because there are 1000 litres in a cubic metre, its density is $1000 \, \text{kg m}^{-3}$. Density is a scalar quantity.

Relative density

The mild steel that we use as a general engineering material in the workshop has a density of $7870 \, \text{kg m}^{-3}$. This means that it is 7.87 times more dense than water. This figure is called its **relative density** and sometimes it is also called its **specific gravity**. Mercury is even more dense than steel. Its density is $13\,600 \, \text{kg m}^{-3}$. This means that it is 13.6 times heavier than water and so its relative density is 13.6.

Did you know?

The platinum kilogram is kept under controlled conditions and closely guarded, not just because platinum is a valuable material, but because it sets the standard against which all devices and machines used for measuring mass must conform. It is called the **international prototype kilogram**. The kilogram is the only fundamental SI unit to be defined in terms of an actual object. Copies of the prototype exist in other countries and periodically they are brought to France to check them against the prototype.

Force

Force is measured in newtons (abbreviation N). A newton is defined as the force needed to accelerate a mass of 1 kg at a rate of $1\ \mathrm{m\,s^{-2}}$ when there is no frictional or other kind of resistance present. If you can imagine a mass of 1 kg floating in space, a steady force of 1 N would cause it to pick up speed at this rate. Force is a vector quantity.

Weight

In scientific and engineering terms the weight (symbol F) of an object is the force that the earth's gravity exerts on it. This varies slightly in different regions, but its accepted value in calculations is 9.81 N pulling down on each kilogram of mass. Mass and weight are often confused by non-scientific people. Strictly speaking, your bathroom scales record your mass in kg and you should multiply this by 9.81 to find your weight in newtons. Sometimes in calculations we use specific weight (symbol w), which is related to density. It is the weight per cubic metre of a substance: i.e. $w = \rho g\ \mathrm{N\,m^{-3}}$.

Moment of a force

When a force is applied to a body it sometimes tries to make it rotate. A force acting on the rim of a wheel or pulley will do this, as in Figure 4.26. The moment of a force is a measure of this turning effect. It is the product of the force and the turning radius that is at right angles to it. Its units are newton-metres (N m).

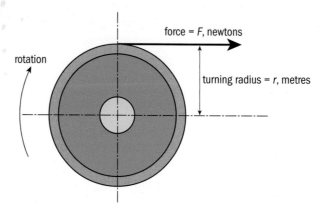

Figure 4.26: Moment of a force on a wheel or pulley

Moment of force = Force × Perpendicular turning radius

Moment of force = Fr (N m)

Displacement

Displacement is the distance moved by a body, measured in a straight line from the starting point to the finish position. The body might not have moved along this straight line, as shown in Figure 4.27.

Displacement is a vector quantity because it has direction. It can be represented by a line drawn to scale on a vector diagram, as in Figure 4.27.

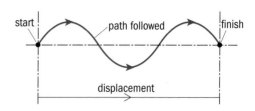

Figure 4.27: Displacement diagram

Velocity

The velocity of a moving body is its rate of change of displacement. Like displacement, it is a vector quantity. Its magnitude is measured in $m\,s^{-1}$ or $km\,h^{-1}$, which is its speed. But the body will also be travelling in some given direction. Very often we omit to mention this, especially if the body is travelling in a straight line, but a full statement of velocity should include both the speed and the direction of travel. As with force and displacement, this enables us to draw the velocity of a body to scale on a vector diagram.

Acceleration

Acceleration is also a vector quantity. It is the rate of change of velocity and its magnitude is usually measured in $m\,s^{-2}$, i.e. the change of speed per second. A body can of course be speeding up or slowing down and this affects the direction in which its acceleration is drawn on a vector diagram. If a body is speeding up, we draw its acceleration vector in the direction of travel. If the body is slowing down, however, its acceleration vector is drawn in the direction opposite to its travel.

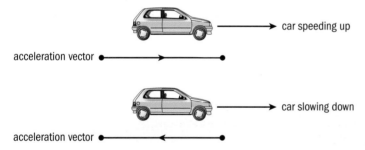

Figure 4.28: Acceleration vectors

Work

When a force acts on a body and causes it to move, work is done. Work is defined as the force multiplied by the distance moved by the body in the direction in which the force acts. The SI unit of work is the joule (J). It is defined as the work done when a force of 1 newton moves a body through a distance of 1 metre along its line of action. Work is a scalar quantity.

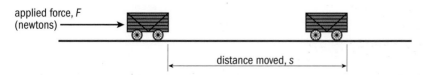

Figure 4.29: Work done

When a force of F newtons moves through a distance of s metres, as in Figure 4.29, the work done is given by

Work done = Force × Distance moved

Work done = Fs joules $\qquad$ (15)

Power

Power is a scalar quantity. It is the rate at which work is done. That is, it is the work done per second. The SI unit of power is the watt (W). It is defined as the work done when a force of 1 newton moves a body through a distance of 1 metre in a time of 1 second. If the time taken to move the body in Figure 4.28 is t seconds, the average power developed will be

Average power = $\dfrac{\text{Work done}}{\text{Time taken}}$

Average power = $\dfrac{Fs}{t}$ watts $\qquad$ (16)

Now the distance s divided by the time t is the average velocity v at which the body shown in Figure 4.28 is travelling and so we can also calculate the average power developed using the formula,

Average power = Force × Average velocity

Average power = Fv (avg) watts $\qquad$ (17)

If the body is speeding up or slowing down, this formula can also be used to calculate the power that is being developed at an instant when the velocity has a value of v ms⁻¹.

Instantaneous power = Force × Instantaneous velocity

Instantaneous power = Fv (inst) watts $\qquad$ (18)

Did you know?

Scalar quantities such as mass, work and power only have magnitude (size). Vector quantities such as force, velocity and acceleration have magnitude and direction.

BTEC Assessment activity 4.3

Read carefully through your notes that define the parameters of static and dynamic mechanical systems:

- mass
- weight
- force
- moment of a force
- density
- relative density
- displacement
- velocity
- acceleration
- work
- power.

Scores:

18–22	Excellent
14–17	Good
10–13	Fair, but more revision needed
Less than 10	A better understanding needed before assessment by your tutor.

Then, without referring to your notes, test yourself by writing down the definition and unit of measurement for each of these parameters and saying whether it is vector or scalar quantity. Check your answers and give yourself 1 point for a correct definition and 1 point for the correct unit of measurement. **P3**

4.2.2 Statics

Parallelogram of forces

When two forces act on a body they may try to pull it in different directions, as shown in Figure 4.30a. This drawing is called a space diagram. It shows accurately the directions in which the forces act, but the forces themselves are not drawn to scale. The combined effect of F_1 and F_2 is called the **resultant *R*** of the system. You can think of it as the one single force that could replace F_1 and F_2 and exert the same pull on the body. For the body to remain stationary, an equal and opposite force called the **equilibrant *E*** must be applied.

We can find the resultant and the equilibrant by drawing a force vector diagram, as in Figure 4.30b. We draw the two forces F_1 and F_2 to a suitable scale from the same point. We then use them to construct a parallelogram, as shown in Figure 4.30b, whose diagonal gives the magnitude and direction of the resultant. The double arrowhead on the resultant (and the equilibrant) is used to show that it is the combined effect of F_1 and F_2.

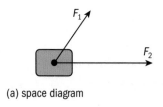

(a) space diagram

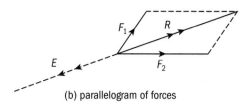

(b) parallelogram of forces

Figure 4.30: Parallelogram of forces

Worked example 4.8

Find the magnitude and direction of the resultant and equilibrant for each of the systems of forces whose space diagrams are shown in Figure 4.31.

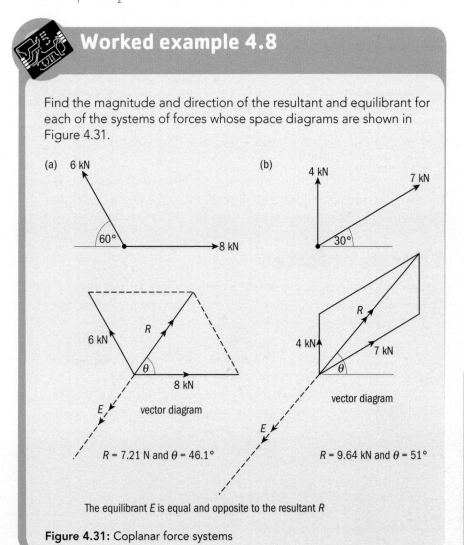

(a) 6 kN

60°

8 kN

6 kN

R

θ

8 kN

E

vector diagram

$R = 7.21$ N and $\theta = 46.1°$

(b) 4 kN

7 kN

30°

R

4 kN

7 kN

θ

E

vector diagram

$R = 9.64$ kN and $\theta = 51°$

The equilibrant *E* is equal and opposite to the resultant *R*

Figure 4.31: Coplanar force systems

Key terms

Resultant – the resultant of a system of forces is their combined pulling effect on a body.

Equilibrant – the equilibrant is the additional force that must be applied equal and opposite to the resultant, to keep the body in a state of equilibrium.

Conditions for static equilibrium

When a body is at completely at rest we say that it is in a state of **static equilibrium**. There are three possible states of static equilibrium, which we can describe by considering the different ways in which a cone can be at rest.

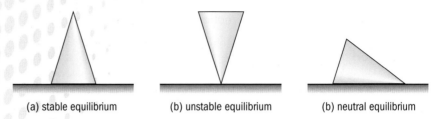

(a) stable equilibrium   (b) unstable equilibrium   (b) neutral equilibrium

Figure 4.32: States of static equilibrium

If the cone in Figure 4.32a is tilted slightly and released, it will return to its stable equilibrium position. The cone in Figure 4.32b is finely balanced on its apex. It is unstable and the slightest force will cause it to topple. The cone in Figure 4.32c is in a state of neutral equilibrium. An applied force will cause it to roll on its side and settle in a new position when the force is removed.

The conditions necessary for all the forms of static equilibrium are as follows:

- The vector sum of the forces acting on a body must be zero so that there is no movement in any direction.

- The sum of moments of the forces acting on a body must be zero so that there is no rotation in any direction

In this unit we shall be considering only bodies acted upon by coplanar forces. That is to say, forces that all act in the same plane, i.e. two-dimensional force systems.

Triangle of forces

When three coplanar forces act on a body and hold it in a state of static equilibrium, as shown in Figure 4.33a, the lines of action of the forces must all pass through a common point.

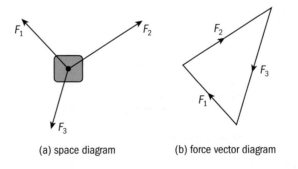

(a) space diagram   (b) force vector diagram

Figure 4.33: Triangle of forces

The forces are said to be **concurrent** and when we draw their force vectors, each added nose to tail as shown in Figure 4.33b, they form a closed triangle. This is called **vector addition**.

Provided that you know the magnitude and direction of two of the forces F_1 and F_2, you can construct the triangle and measure the magnitude and direction of the third force F_3. Similarly, if you know the magnitude and direction of only one of the forces F_1 together with the directions of the other two, you can still construct the triangle and measure the magnitude of F_2 and F_3.

It is worth remembering the fact that a body can be in static equilibrium under the action of three coplanar forces only if the forces act at a single point, i.e. when the forces are concurrent. We shall be considering only concurrent coplanar forces in this unit, but this fact may come in useful when you progress to more advanced work.

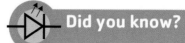

Did you know?

In **vector addition**, the vectors are added nose to tail on a vector diagram.

Worked example 4.9

If the body shown in Figure 4.34 is in a state of static equilibrium, find the angle θ that the 7 kN force must make with the horizontal.

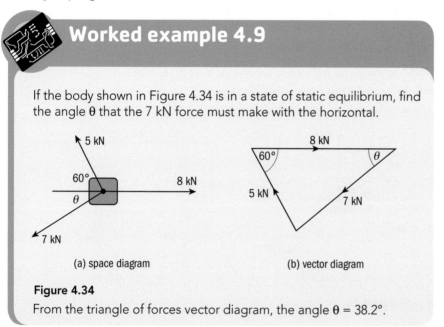

(a) space diagram   (b) vector diagram

Figure 4.34
From the triangle of forces vector diagram, the angle θ = 38.2°.

Polygon of forces

Now consider a body acted on by three or more concurrent coplanar forces in such a way that it is *not* in equilibrium.

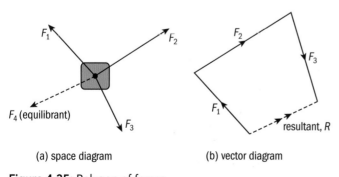

(a) space diagram   (b) vector diagram

Figure 4.35: Polygon of forces

When the force vectors are added together, the vector diagram (Figure 4.35) is not a closed triangle. There is a gap between the tail of F_1 and the nose of F_3. The line joining these, which is shown dotted in Figure 4.35b, gives the magnitude and direction of the resultant R of the three forces and if the body is not secured it will be pulled in this direction. The additional force F_4 that must be added to the system to maintain static equilibrium is equal and opposite to the resultant, as shown in Figure 4.35a. This is of course the equilibrant of the system.

You can use this method of solution to find the resultant and equilibrant of any number of concurrent coplanar forces. It is just that your polygon of forces will have more sides. It is good practice to work clockwise around the body when adding the force vectors together, as in Figure 4.35b. Remember to always add them nose to tail so that the force vectors chase each other around the polygon, as indicated by the arrowheads.

Did you know?

When a system of concurrent coplanar forces is in equilibrium, the force vector diagram is a closed polygon.

Worked example 4.10

Determine the magnitude and direction of the resultant and equilibrant of the coplanar force system shown in Figure 4.36.

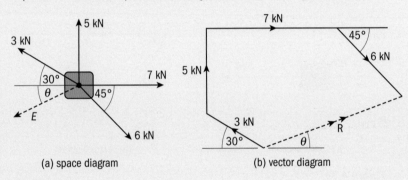

(a) space diagram   (b) vector diagram

Figure 4.36

From the vector diagram $R = 8.93$ kN and it makes an angle $\theta = 14.7°$ to the horizontal, as shown. The equilibrant E is equal and opposite to the resultant R.

Principle of moments

One of the conditions for static equilibrium is that the sum of moments of the forces acting on a body must be zero, so that there is no rotation clockwise or anticlockwise. This is sometimes called the **principle of moments**. For static equilibrium we can write it as:

Clockwise moments = Anticlockwise moments

We can use the principle to find the support reactions of a simply supported beam, as shown in Worked Example 4.11. The load that acts on a support of any kind is said to be an **active force**. If the load and its support are in static equilibrium we say that the support exerts an equal and opposite **reactive force**, which is why we use the term **support reaction**.

Worked example 4.11

Determine the load carried by each support of the simply supported beam shown in Figure 4.37.

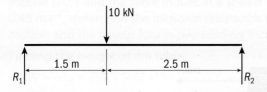

Figure 4.37

Begin by equating moments about the right-hand support to find the reaction R_1:

Clockwise moments = Anticlockwise moments

$$R_1 \times 4 = 10 \times 2.5$$

$$R_1 = \frac{10 \times 2.5}{4}$$

$$R_1 = 6.25 \text{ kN}$$

Now equate moments about the left hand support to find the reaction R_2.

Clockwise moments = Anticlockwise moments

$$10 \times 1.5 = R_2 \times 4$$

$$\frac{10 \times 1.5}{4} = R_2$$

$$R_2 = 3.75 \text{ kN}$$

Finally, check that the two reactions that you have calculated add up to the downward load of 10 kN:

$$R_1 + R_2 = 6.25 + 3.75 = 10 \text{ kN}$$

i.e. the answers are correct.

Friction

Frictional resistance is always present in mechanical systems. Sometimes it is an essential part of the system, as in friction clutches and brakes. In other instances, such as in bearings and gear trains, it wastes energy and can lead to overheating and breakdown. Efficient lubrication systems are needed for these components.

Within limits, the frictional resistance between dry surfaces in sliding contact is found to be dependent on two things and independent of two other things. The following statements are sometimes known as Coulomb's laws of friction. (This is the same Coulomb that the unit of electric charge is named after.)

- Frictional resistance depends on the normal force or pressure between the surfaces. (Here the word 'normal' means perpendicular to, or at right angles to, the surfaces.)

- Frictional resistance depends on the roughness of the two surfaces.

BTEC Assessment activity 4.4

P4 **P5** **P6** **M2** **D2**

1. Draw the force vector diagram for the system of concurrent coplanar forces shown in Figure 4.43. Use it to determine the resultant and the equilibrant of the system. **P4**

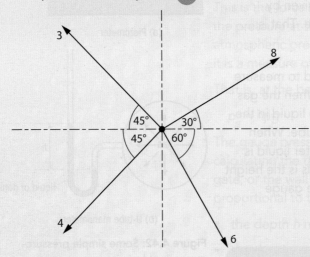

Figure 4.43:

2. Describe the conditions that must apply for a body to be in a state of static equilibrium. **M2**

3. A commuter train starts from rest at a station and is uniformly accelerated for 2 minutes, during which it travels a distance of 0.5 km. It then continues at a steady velocity for 2.5 km before being uniformly retarded to rest at the next station over a distance of 0.75 km.

Calculate:

(a) the uniform acceleration during the first 2 minutes of the journey.

(b) the velocity at the end of the acceleration period

(c) the time for which the train is travelling at a steady velocity

(d) the retardation as the train is coming to rest

(e) the total time taken in travelling between the stations. **P5**

4. An electric truck pulls a girder of mass 450 kg along a horizontal floor at a steady speed of 0.5 m s⁻¹ for a period of 12 s. The coefficient of kinetic friction between the girder and floor is 0.4 and the towrope is parallel to the floor.

Calculate:

(a) the distance travelled

(b) the force exerted by the truck

(c) the work done

(d) the power dissipated **D2**

5. A tank containing oil of relative density 0.85 has a circular cover of diameter 600 mm in its base. Calculate the gauge pressure at the base and the force acting on the cover, when the depth of oil contained is 3.5 m. What will be the absolute pressure at the base of the tank if atmospheric pressure is 1.013 bar? **P6**

Grading tips

A scale of 10 mm = 1 kN will be ideal for the vector diagram in Task 2; remember to add the force vectors nose to tail. In Task 3 you may give a practical example to support your description.

Work tidily and set out your calculations in a logical order. Make full use of your electronic calculator and always carry out a calculation a second time to confirm your answer. If there is an alternative formula for the parameter you have calculated, use it as a check.

PLTS

Analysing information and solving problems relating to pressure at depth will help you to develop your **independent enquirer** skills in handling data.

Functional skills

Selecting and applying formulae to solve problems on pressure at depth and checking your answers will help you to develop your **mathematical** skills.

WorkSpace Ann Marjoram

Aero engineer

On leaving school I obtained an apprenticeship with a leading aerospace company and studied part-time BTEC courses at our local college of further education. I achieved good results and my firm sponsored me to study for a degree in mechatronics. This is a combination of mechanical engineering, electrical engineering and IT. I am now employed as a systems engineer working as part of a development team on electro-mechanical systems.

Typical day

In a typical day I might have meetings with colleagues to discuss the progress of new systems and problems that have arisen in their design and testing.

In the past, these systems have ranged from heating and cooling systems to the operation of flight deck controls and instrumentation. I might then turn my attention to the development and testing of a particular system. This often involves calculations and the interpretation of test results. I often have to call other firms for expert advice and arrange for tests to be carried out by our development technicians.

It sometimes surprises me how many of the basic formulae that I learned at college I still use almost every day. When you are learning, you realise that they are building blocks, but you are not quite sure what the final building will be. As you progress, you begin to see how the basic laws and principles underpin all the complex systems that we deal with.

The best thing about the job

The best thing I like about the job is being trusted to get on with a problem and come up with a solution. That said, I also enjoy being part of a team with more experienced engineers who I can turn to for advice and suggestions. I enjoy air travel and often think that thanks to our efforts it is one of the safest forms of transport.

Think about it!

1. What areas have you covered in this unit that could provide you with the background knowledge and skills for a career in aerospace engineering? Write a list and discuss it with your peers.

2. Working in small groups, draw up a list of products whose design requires knowledge of both mechanical and electrical engineering principles.

Check your understanding

1. What are the units of electric current, potential difference and electrical resistance?

2. State and explain Ohm's law.

3. What does Fleming's left-hand rule tell you?

4. What is the formula for calculating the force on a current-carrying conductor in a magnetic field?

5. What is the difference between mass and weight?

6. What are the conditions for a body to be in a state of static equilibrium?

7. What does the area under a graph of velocity against time tell you?

8. How do you calculate the pressure at a depth below the surface of a liquid?

edexcel **:::**

Top tips

- Keep a tidy file of notes and worked examples that you can refer to when revising for an assessment.

- Recognise the areas where you are least confident and don't hesitate to seek help from your tutors.

- Carefully read your assignment briefs and make sure that you have addressed everything you have been asked to do.

- It is good practice to word-process your assignment work. It will develop your IT skills and make it easier for your tutor to assess your work and give you feedback.

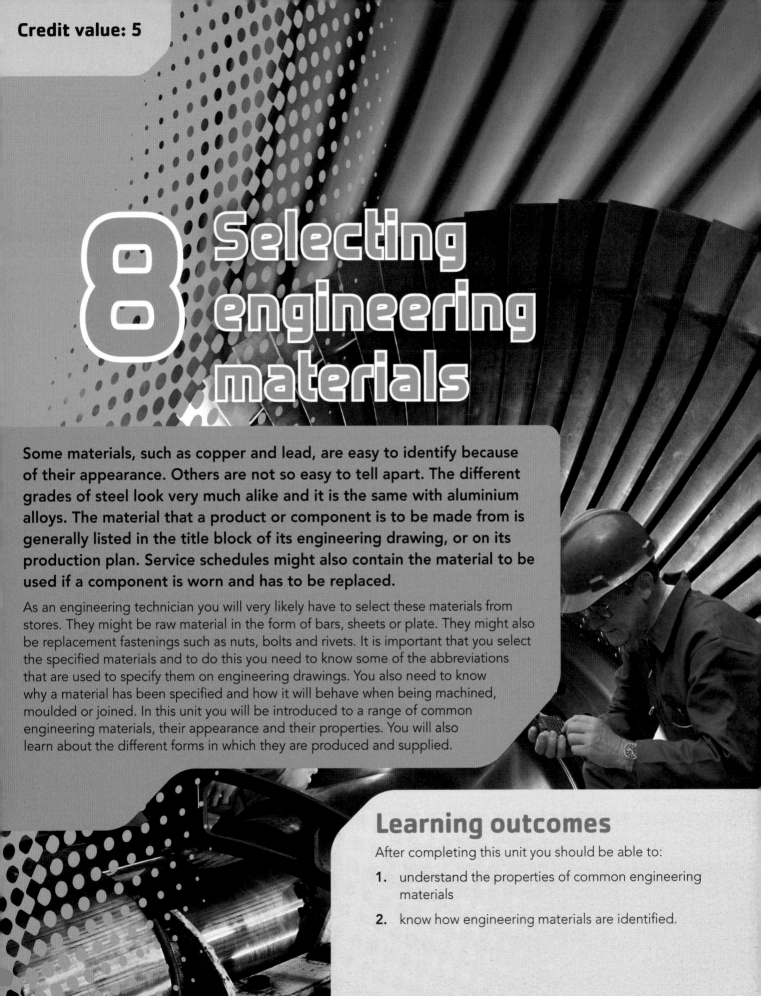

8 Selecting engineering materials

Some materials, such as copper and lead, are easy to identify because of their appearance. Others are not so easy to tell apart. The different grades of steel look very much alike and it is the same with aluminium alloys. The material that a product or component is to be made from is generally listed in the title block of its engineering drawing, or on its production plan. Service schedules might also contain the material to be used if a component is worn and has to be replaced.

As an engineering technician you will very likely have to select these materials from stores. They might be raw material in the form of bars, sheets or plate. They might also be replacement fastenings such as nuts, bolts and rivets. It is important that you select the specified materials and to do this you need to know some of the abbreviations that are used to specify them on engineering drawings. You also need to know why a material has been specified and how it will behave when being machined, moulded or joined. In this unit you will be introduced to a range of common engineering materials, their appearance and their properties. You will also learn about the different forms in which they are produced and supplied.

Learning outcomes

After completing this unit you should be able to:

1. understand the properties of common engineering materials

2. know how engineering materials are identified.

Assessment and grading criteria

This table shows you what you must do in order to achieve a **pass**, **merit** or **distinction** and where you can find activities in this book to help you.

To achieve a **pass** grade the evidence must show that you are able to:	To achieve a **merit** grade the evidence must show that, in addition to the pass criteria, you are able to:	To achieve a **distinction** grade the evidence must show that, in addition to the pass and merit criteria, you are able to:
P1 describe the properties that are used to define the behaviour of common engineering materials **Assessment activity 8.1 page 144**	**M1** explain the choice of material for a given engineering component based on the material's properties **Assessment activity 8.1 page 144**	**D1** establish that a material has the required properties for a given application. **Assessment activity 8.2 page 150**
P2 review the properties of a given ferrous metal and a given non-ferrous metal **Assessment activity 8.1 page 144**	**M2** select an appropriate form of supply for a given material requirement. **Assessment activity 8.2 page 150**	
P3 review the properties of a given organic material, a given thermoplastic, a given thermosetting polymer and a given smart material **Assessment activity 8.1 page 144**		
P4 identify symbols and abbreviations used on engineering documentation **Assessment activity 8.2 page 150**		
P5 identify the forms of supply available for a given engineering material. **Assessment activity 8.2 page 150**		

How you will be assessed

This unit will be assessed by means of investigative assignments and/or through the responses given to engineering problems and questions that cover the requirements of the assessment criteria.

Your assessment will most likely be in the form of a number of written assignments in which you will need to show an understanding of the properties, uses and forms of supply of a range of engineering materials. In certain cases your tutor may choose to assess your understanding through oral questions.

Mark, 18-year-old apprentice

Because I had to change schools partway through my final year, my GCSE results weren't very good. The subjects I enjoyed doing most were CDT and IT. I wanted to follow a practical career, but my job applications were unsuccessful. My careers teacher suggested that I should think about doing a BTEC Level 2 First in Engineering at our local further education college, as this would be a good way into an engineering career.

I enrolled on the course and found it to be just what I wanted. I particularly enjoyed the IT and workshop periods. The college is very well equipped and it was interesting to see how three-dimensional modelling and computer-aided machining are done. Partway through the course I was advised of a training vacancy at an engineering firm. I applied for the job and was successful. I now work with a team of maintenance technicians who service the plant and equipment and deal with any emergencies that arise.

I completed the course with grades that were mostly merits and distinctions and I have now moved on to the BTEC Level 2 National Course in Operations and Maintenance Engineering. I think that the BTEC Level 2 First in Engineering gives a good introduction to engineering and I would like to have had the option to do it at school.

Over to you

- Which section of the unit do you think you will find the most interesting?
- What area of this unit do you think you will find the most useful?
- How can you best prepare for unit assessment?

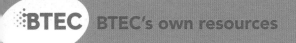

8.1 Understand the properties of common engineering materials

Start up

Have a look at a simple electrical appliance, such as a toaster or kettle and think about the materials which the various parts are made from. Why did the designer of the product select a particular material for each component?

The reason is that the properties of the material have been matched to the operating requirements of the component. For example the outside casing of the toaster will be made from a material which has good surface finish properties while the heating elements will be made from a material which has excellent heat resisting properties.

Working with a colleague, discuss the properties of the materials which are used for the surface of the screen display, the body casing and the battery contacts of a mobile phone.

Some metals, such as gold, silver, lead, tin and copper, are almost pure when found in their natural state. They were known and used in very ancient times. You will have heard about the Bronze Age (3300–1200 BC). That was when it was discovered that mixing copper and tin gave bronze, which is much stronger and hard-wearing than either of its ingredients. A little later iron was discovered. These metals, along with wood, bone, stone and clay, were the only engineering materials available until fairly modern times. Now we have a wide range of metals, plastics and ceramics at our disposal, some of which we shall be looking at in detail.

8.1.1 Material properties

We need to have a way of describing how metals behave or react in service. This is what we call their properties. A material can have mechanical **properties**, electromagnetic properties and chemical/durable properties. Some of these are as follows.

Mechanical properties

These include tensile strength, hardness, toughness and brittleness, malleability and ductility.

- **Tensile strength:** The ultimate tensile strength (UTS) of a material is the maximum load that each unit of cross-sectional area can carry before it fails. We also call this the tensile stress at failure. We measure it in newtons (N) carried per square metre (m²). The SI name for this unit is the pascal (Pa). The pascal is a small unit and so we often have to use megapascals (MPa). 1 MPa = 1 000 000 Pa.

Key term

Properties – the qualities or power that a substance has.

- **Hardness:** The hardness of a material is its resistance to wear, abrasion or indentation. A hard material is difficult to cut and very hard materials often have to be cut and polished by grinding.

- **Toughness and brittleness:** The toughness of a material is its resistance to sudden impact and shock loading. Tough materials are able to absorb the energy when something strikes them without breaking. The opposite of toughness is brittleness. Glass and the pottery that we use for cups and plates are quite hard and wear resistant, but are very brittle and easily shattered.

- **Malleability and ductility:** A material is malleable when it can easily be pressed or forged into shape. Some metals are malleable when cold, but others have to be heated to make them malleable. Rivets have to be malleable so that their heads can be formed to shape and so does the sheet metal used to make car body panels. Do not confuse malleability with ductility. Ductility is the ability of a material to be pulled or drawn out in length to make thin rods and wire. Some materials are malleable but not very ductile. Some have both properties and behave rather like chewing gum. Another way of looking at these two properties is to say that malleable materials can be formed to shape using mostly compressive forces, whereas ductile materials can be formed to shape using mostly tensile forces.

Electromagnetic properties

These include electrical conductivity and ferromagnetism.

- **Electrical conductivity:** This is the ability of a material to conduct an electric current. Most metals are good conductors, whereas plastics and ceramics are bad conductors and are used as insulation materials.

- **Ferromagnetism:** Metals that contain large amounts of iron, nickel and cobalt can generally be made magnetic. They are used to make permanent magnets and electromagnets. The Latin name for iron is *ferrum* and these metals are said to be **ferromagnetic**, or to possess the property of ferromagnetism.

Chemical and durability properties

These include corrosion resistance, solvent resistance and environmental durability.

- **Corrosion resistance:** Some metals go rusty and eventually they will corrode away completely. A chemical reaction occurs between the metal, the oxygen in the air and moisture. Others are not affected very much and we say that they have high corrosion resistance. Gold, for instance, can lie buried for hundreds of years and still be bright and shiny when it is dug up.

Did you know?

When a magnet is heated, a point is reached at around 800°C where its magnetism disappears. It is called the *Curie temperature* of the material.

- **Solvent resistance:** Some rubbers and plastics are attacked by certain chemicals. We call these chemicals *solvents*. Other materials are very stable and are not affected. We say that they have a high solvent resistance. Petrol, diesel oil and lubricating oils can act as solvents and we have to be careful to select rubbers and plastics with a high solvent resistance if they are going be in contact with these substances.

- **Environmental degradation:** If left unprotected, wood will rot when exposed to moisture and certain plastic materials become very brittle when exposed to sunlight over a long period. We say that they undergo environmental degradation. Wood becomes more durable when painted or treated with chemicals. Plastics, such as the kind used for spouts and guttering on your house, tend to stay flexible for longer if coloured black – as many of them are.

- **Wear resistance:** We have already mentioned the property of hardness, which by definition is wear resistance. This can also be thought of as a durability property. Machine components that slide or roll over each other, such as gear teeth, need to be hard and durable.

Activity: Defining material properties

Read carefully through your notes that define the properties of engineering materials. They are:

- tensile strength
- hardness
- toughness and brittleness
- malleability and ductility
- electrical conductivity
- ferromagnetism
- corrosion resistance
- solvent resistance
- resistance to environmental degradation.

Without referring to your notes, test yourself by writing down the definition for each of these properties. Check your answers and give yourself 1 point for a correct definition.

Scores:

8–9 Excellent

7–8 Good

6–7 Fair, but more revision needed

Less than 6 A better understanding needed before assessment by your tutor.

8.1.2 Common engineering materials

We use a wide range of materials in modern engineering. They include metals, plastics, ceramics and hard and soft woods. Many of these have been developed to serve a particular purpose. This is especially true of a range called **smart materials**, whose properties can change very quickly in response to pressure, temperature or the presence of a magnetic or electrostatic field.

Ferrous materials

Iron is the main constituent of **ferrous materials**. In its pure form, iron is a soft grey metal that doesn't cast very well when molten and doesn't give a good surface finish when machined. However, the addition of small amounts of carbon greatly improves its properties, giving us a range of cast irons and steels.

Metals are made up of crystals that are sometimes also called **grains**. Sometimes these can be seen with the naked eye, but more often they have to be viewed under a microscope.

- **Cast iron:** In cast iron the carbon content is around 2–4%. This makes it very fluid when molten and enables it to be cast into complicated shapes. This property gives it its name. When looked at under a microscope it can be seen to contain flakes of carbon in the form of graphite. This is the same material that pencil leads are made from. Not only does carbon make the iron easier to cast, it also makes it easier to machine and the graphite flakes act as a lubricant when two pieces of cast iron rub together. We say that cast iron is **self-lubricating**.

  Cast iron, or 'grey cast iron' as it is often called, is widely used for engine parts, valve bodies, lathe beds and decorative garden furniture. Perhaps its main disadvantage is that, unless specially treated, it tends to be brittle and should not be subjected to high tensile forces. We say that it is strong in compression but weak in tension.

- **Mild steel:** Like cast iron, steel is a mixture of iron and carbon. The difference is that the carbon is present in much smaller, controlled amounts. In mild steel this is in the range 0.1–0.3%. Mild steel is easy to machine: it has good tensile strength and a fair degree of malleability and ductility when cold-worked. When heated to a bright red colour it becomes much more malleable and ductile and can be more easily pressed, forged and rolled into shape.

  Mild steel is the most common engineering material that we use. It is used for girders, pipes, ships' hulls, gates and railings and for general workshop purposes. Mild steel components are sometimes given a hard surface layer by a process called **case hardening**. This gives

Key term

Ferrous material – a material that is composed of or contains iron.

The body of this motor car is made of mild steel which has been pressed to form various contours.

them a wear-resistant surface while retaining the strong and tough core. One method of doing this is to pack mild steel components in a cast iron box containing a carbon-bearing material such as graphite. When heated to a high temperature the carbon soaks into the steel giving it an outer case with a high carbon content.

- **Medium carbon steel:** Medium carbon steel contains 0.3–0.8% carbon. It is stronger and tougher than mild steel and a little more difficult to machine. Its hardness and toughness can be further increased by the heat treatment processes called **hardening and tempering**. Medium carbon steel is used for hammers, chisels, punches, couplings, gears and other engineering components that have to be wear and impact resistant.

- **High carbon steel:** High carbon steel contains 0.8–1.4% carbon and can be made very hard and tough by hardening and tempering. It is used for sharp-edged cutting tools such as wood chisels, files, screw-cutting taps and dies and craft knife blades. Mild, medium and high carbon steels are sometimes called **plain carbon steels** because iron and carbon are their main constituents.

- **Stainless steel:** In addition to iron and carbon, stainless steel contains chromium and nickel. It belongs to a class of ferrous metals called **alloy steels**. The added ingredients make it corrosion resistant and also very tough. It is widely used in food preparation areas, for surgical instruments, for cutlery and very often for ornaments and decorative trim.

See the Appendix for further information on ferrous materials.

Non-ferrous metals

Some non-ferrous metals are widely used in engineering in their almost pure form. Unlike most ferrous metals they are generally corrosion resistant.

- **Aluminium:** Aluminium is probably the most widely used of the non-ferrous metals and is one of the lightest. It is malleable, corrosion resistant and a good conductor of both heat and electricity. It is widely used for cooking utensils and in overhead power transmission cables, where a core of high-tensile steel is surrounded by aluminium conductors. In its pure form it has low tensile strength when compared with steel, but when alloyed with small amounts of copper, silicon, iron, magnesium and manganese its properties are greatly improved. Aluminium alloys are widely used in the aircraft and motor industries.

- **Lead:** Lead has been used since ancient times. It is a heavy grey metal that is very malleable and highly corrosion resistant. It has low tensile strength and is used in its pure form as a roofing material and as a lining for tanks containing chemicals. When lead is mixed with tin it produces the range of alloys called soft solders. Soft solder is used for joining copper, brass and mild steel components and for making electrical joints.

 Did you know?

An alloy is a mixture of metals. It may also be a mixture of a metal and a non-metal, so long as the final product has metallic properties. Cast iron and plain carbon steels are really alloys, but we don't usually refer to them as such.

 Did you know?

Many industries have been pushed by legislation to find alternatives to lead. All plumbing joints must use lead free solder, and most of the electronics industry is following this.

- **Copper:** Copper, with its characteristic red colour, is malleable, ductile, corrosion resistant and an excellent conductor of heat and electricity. It is used in its almost pure form for water pipes, for wires and cables for conducting electricity and for cooking utensils. Copper is not as strong as steel, but it has fairly good tensile strength, which can be improved by alloying it with tin to make bronze and with zinc to make brass.

- **Tin:** Tin is a soft and malleable metal that is highly corrosion resistant. In its pure form it is used as a protective coating for mild steel, which is then known as tinplate. This is used for the tins (or cans) in which you buy soups, beans and so on. Tin is also used with lead to make solders and alloyed with copper to make bronzes.

- **Zinc:** Zinc is a soft but rather brittle material. Like most of the other non-ferrous metals it is highly corrosion resistant. It is used as a coating for mild steel, which is then said to be galvanised. Galvanised steel is used for waste bins, metal buckets and building materials. It is easy to recognise from the feathery pattern that the zinc forms on the steel surface.

Table 8.1 shows the tensile strengths, densities and melting points of some common ferrous and non-ferrous metals.

Metal	Ultimate tensile strength (MPa)	Density (kg/m³)	Melting point (°C)
Cast iron	200	7200	1200
Mild steel	500	7800	1500
Medium carbon steel	750	7800	1450
High carbon steel	900	7800	1400
Copper	230	8900	1083
Zinc	200	7100	420
Tin	Very low	7300	230
Aluminium	95	2700	660
Lead	Very low	11 300	327

Table 8.1: Properties of some common metals

- **Brass:** Brass is an alloy of copper and zinc. The proportions vary from 85% copper and 15% zinc to 60% copper and 40% zinc, depending on the material's final use. A high copper content makes the brass ductile and suitable for deep drawing into tubes and cartridge cases. A high zinc content makes the brass more fluid when molten and better suited for making castings.

- **Bronze:** Also known as tin bronze, this is an alloy of copper and tin. The proportions vary from 96% copper and 4% tin to 78% copper

Activity

Use the internet to find images of the materials you have read about in this unit. Try to find a few products of each.

Did you know?

When tinplate is scored through the tin coating, the steel beneath will begin to rust. But when the zinc coating of galvanised steel is scored through, the steel does not rust. See if you can find out why.

and 22% tin, depending on its final use. The effect of the tin is similar to that of the zinc in brass. A high copper content makes the bronze malleable and ductile, with good elasticity after forming to shape while cold. A high tin content makes the bronze more fluid when molten and better suited for making castings. Bronze with a high copper content is used for electrical contacts and instrument parts. Bronze with a high tin content is used for casting into pump and valve components.

- **Aluminium bronze:** This is a copper–aluminium alloy with an aluminium content of 5–10%. It is used for boiler and condenser tubes, chemical plant components and boat propellers. As with the addition of zinc and tin to copper, the addition of aluminium gives an alloy that is much stronger that its constituents and more fluid when molten.

- **Aluminium alloys:** A wide range of aluminium alloys have been developed over the years. The addition of very small amounts of silicon, copper, magnesium and manganese greatly increases the strength of aluminium. One of the most widely used aluminium alloys is **duralumin**, which contains 4% copper and 1% magnesium and is almost as strong as mild steel. It is used for aircraft parts and for motor vehicle panels. Other aluminium alloys have been developed for casting and some of them may be hardened by heat treatment.

See the Appendix for further information on non-ferrous materials.

Did you know?

The 'copper' coins in your pocket are in fact a kind of tin-bronze if made before 1992. Since that date they have been made from copper coated steel and can be picked up by a magnet. The 'silver' coins are made from an alloy of copper and nickel called cupro-nickel.

This aeroplane is made largely from aluminium alloys

Thermoplastics

Plastics and rubbers are classed as **polymer materials**, or just **polymers** for short. Polymers are long intertwined chains of molecules, rather like spaghetti. They basically consist of hydrogen and carbon atoms with others attached, which gives them their different properties.

The word 'plastic' means malleable and ductile, but this is not a very good description for some polymer materials, which can be quite hard and elastic. The description is more fitting at high temperatures, where a group of polymer materials called **thermoplastics** become soft and easy to mould. Some of the more common thermoplastics are listed below. Don't worry about the complicated chemical names. You need only remember the everyday names that we give them.

- **Polyesters:** One of the best-known polyesters is **Terylene**, which can be drawn out into thin fibres. These are very strong and hard wearing and they are used in textiles for clothing. Polyester fabric is also used as a reinforcement for rubber in drive belts, rubber tyres and hoses.

- **Polyamide:** You know this better as **nylon**, which can also be drawn out into a strong thin fibre used as bristles for brushes and for fishing lines. Nylon is also very tough and flexible. It is used in engineering for moulded gears, cams and bearings.

- **Polyethene:** In everyday use this is shortened to **polythene**. It is tough and flexible. We use it as a thin film for wrapping and packaging, but it can also be used for squeeze containers, pipes and mouldings.

- **Polypropene:** This thermoplastic has high strength, good flexibility and a high melting point. It can be moulded into kitchen utensils, tubes and pipes. It can also be produced as a fibre for making ropes. Polypropene pipe is being used in place of copper in domestic central heating systems.

- **Polychloroethene:** We know this better as **polyvinyl chloride** or **PVC**. It can be made hard and tough, or soft and flexible. When hard it is used for window frames, drain pipes and guttering. When soft it is used for the different colours of insulation around electrical wiring and cables.

- **Polyphenylethene:** This is what we generally know as **polystyrene**. It can be made as a foam for packaging and disposable drinking cups. It can also be made into hard, tough and rigid mouldings such as are used in refrigerator interiors.

- **Methyl-2 methylpropenoate:** This is generally known by its trade name, **Perspex**. Perspex is strong, rigid and transparent. It is used for lenses, aircraft windows and cockpit canopies and also for protective shields and guards on workshop machinery.

- **Polytetrofluoroethene:** This is shortened to **PTFE** and is also known as **Teflon**. It is tough, heat resistant and highly solvent resistant. It has a very smooth surface with a low coefficient of friction making it ideal as a bearing material. Teflon is also used for the non-stick coating on cooking utensils and in sheet or tape form for seals and gaskets.

See the Appendix for further information on thermoplastic materials.

This Dyson vacuum cleaner is made almost entirely of thermoplastics and thermosets

Thermosetting plastics

During their forming process thermosetting plastics, or **thermosets** as they are often called, undergo a chemical change while they are being moulded into shape under the effects of heat and pressure. The polymers become cross-linked together and once these cross-links are formed, they cannot be broken. Thermosets are generally harder and more rigid than thermoplastics and cannot be softened by re-heating.

Thermosetting plastics begin the moulding process in either liquid or powder form. Very often other substances known as fillers are added to improve their mechanical properties. These include very fine sawdust known as wood flour, shredded paper and textiles, glass fibres and carbon fibres. Some of the more common thermosets are as follows.

- **Phenolic resin:** This is more commonly called **Bakelite**. It was one of the earliest thermosetting plastics to be developed. It is hard, heat resistant and solvent resistant. It is a good electrical insulator and can be machined. The colour of Bakelite is limited to brown and black, which rather limits its use for decorative purposes. It is used for electrical components and also for heat-resistant handles such as on kettles and saucepans.

- **Urea-methanal resin:** You probably know this as **Formica**. It has similar properties to Bakelite. It is naturally transparent, but can be produced in a range of colours. Kitchen worktops and worktable tops are often covered with a layer of hard-wearing Formica laminate. It is also used for items of kitchenware, toilet seats and electrical fittings.

- **Methanal-melamine resin:** We call this **melamine** for short. It has similar properties to Bakelite and Formica, but is harder and more resistant to heat. It can also be moulded with a smoother surface finish. The white electrical plugs and sockets in your home are probably moulded from melamine. It is used for other items of electrical equipment, control knobs and heat-resistant handles.

- **Epoxy resins:** These are generally formed by mixing together the liquid resin and a hardener. This triggers the cross-linking process and the resin can then be poured into moulds, or applied with **glass fibre**, **carbon fibre** or **Kevlar** reinforcement matting to the surface of a mould. This is how canoe and boat hulls are made. Reinforced epoxy resins are strong and tough and have good electrical resistance.
  You may hear the term **GRP** stated in relation to the above products. This stands for 'glass-reinforced plastic'. Kevlar is a synthetic fibre. It is an extremely strong material, with a tensile strength said to be five times that of steel of an equal weight. It is widely used by the military for the reinforcement in protective helmets and body armour.
  Epoxy resin and its hardener are sold in tubes for mixing together as an adhesive. One of its trade names is **Araldite**. It makes an excellent adhesive for joining polymer materials and particularly for bonding together the layers that go together to make plastic laminates.

- **Polyester resins:** These have similar properties to epoxy resins, with good heat resistance and a hard wearing surface. They are moulded with the same glass fibre, carbon fibre and Kevlar reinforcement materials. Boat hulls, skis, motor panels, aircraft parts and fishing rods are all moulded using these reinforced resins.

See the Appendix for further information on thermosetting plastic materials.

Organic materials

By organic materials we mean materials with an animal or vegetable origin. Leather, bone and sinew from animals have all been used for engineering purposes in the past and leather still has its uses today. Timber and certain fibrous materials such as jute for making ropes are of vegetable origin. Timber and wood composites are widely used in both engineering and building and some of the more common types are as follows.

- **Hardwoods:** These generally come from broad-leaves trees such as oak, elm, teak and mahogany, which shed their seeds in a seed case. Hardwoods are not necessarily harder than softwoods, although most of them are. They tend to be more dense than softwoods and have shorter fibres. Hardwoods are used for a variety of structural and decorative purposes, as shown in Table 8.2. Because of their cost they are often cut into thin veneers and used as a facing for less expensive timbers.

Did you know?

The very earliest plastic was developed in the USA in the mid 19th century and used instead of ivory for billiard balls. Unfortunately it was unstable and tended to explode on impact, causing people to reach for their guns.

Table 8.2: Hardwoods

Hardwood	Density (kg/m³)	Moisture content (%)	Uses
Elm	550	12	Durable under water. Used for lock gates, piles and outdoor cladding
Oak	720	12	Best wood for structural purposes. Also used for furniture and fittings
Mahogany	720	12	Used for high-quality furniture
Ash	810	12	Used for vehicle bodies and tool handles and shafts
Teak	900	11	Used mainly for indoor and outdoor quality furniture

- **Softwoods:** These come mainly from trees with narrow, spiky leaves and longer fibres, such as conifers. About 80% of the timber used in the world is softwood. Some of the more common softwoods are pine, fir, larch and spruce. Unlike hardwoods, softwoods are relatively quick growing. The are forested and replanted in a planned sequence to conserve stocks and hence support the environment. They are used for furniture and a variety of structural uses, as shown in Table 8.3. Wood is a valuable resource and the waste from sawmills and inferior pieces are recycled in wood composites.

Table 8.3: Softwoods

Softwood	Density (kg/m³)	Moisture content (%)	Uses
Spruce	420	13	Used for boxes, crates, packaging and general construction work
Scots pine	510	12	Used for furniture, flooring and general construction work
Douglas fir	530	12	Used for plywood and heavy construction work
Larch	810	12	Used for outdoor purposes, mining and general purposes

- **Wood composites:** These fall into three general classes: laminated boards, particle boards and fibre boards. **Plywood** is the main type of laminated board. It consists of thin layers of wood bonded together, with their grain directions running alternately at right angles. This reduces the possibility of warping. Marine-quality plywood is specially treated to make it suitable for outdoor use. **Blockboard** is another type of laminate: It consists of strips of wood bonded together and sandwiched between two thin outer sheets. See Figure 8.1.

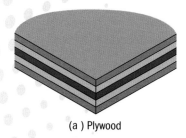
(a) Plywood

(b) Blockboard

(c) Laminated Chipboard

Figure 8.1: Wood composites

- Particle boards are made from bonded wood particles that originate from recycled waste such as sawdust, wood shavings and substandard pieces that have been reduced to particles and fibres. These are bonded with resin to form **chipboard**, which is also sometimes sandwiched between thin outer layers that give an appearance of wood grain. Laminated and veneered chipboard is widely used in the manufacture of furniture and kitchen units.

- Fibre boards are made from fine compressed wood fibres of differing size. They include **hardboard** and **MDF** (medium density fibreboard). MDF is manufactured by a dry process at a lower temperature than hardboard and with a different bonding agent. They are both used extensively in building and engineering.

Smart materials

You may be familiar with spectacle lenses that become darker in sunlight. This is an example of a smart material – that is, a material that can undergo a property change when there is a change in its working environment. Other smart materials are as follows.

- **Piezoelectric materials:** With certain crystals, such as quartz, the application of a force causes a potential difference to be set up across faces at right angles to the force. This is called the **piezoelectric effect**. It is put to use in pressure sensors, vibration recorders and microphones. The same can also occur in reverse where an applied voltage produces a stress in the material. This can cause it to twist or bend by a controlled amount. Smart materials with this kind of response are being applied in aircraft control systems.

- **Shape memory alloys:** These are sometimes called **memory metals**. When deformed they will return to their original shape when heated, or when the forces are removed. They are alloys containing special combinations of copper, zinc, nickel, aluminium and titanium. Shape memory alloys have been used for vascular stents that are placed in restricted blood vessels. Once they are in position, the body's temperature causes them to enlarge, opening up the blood vessels and giving improved blood flow. Shape memory alloys are being developed that will return to their original shape when a magnetic field is applied and it is thought that these will have a faster response than the heat-sensitive types.

- **Magneto-rheostatic fluids:** These fluids contain microscopic magnetic particles suspended in a type of oil. In the presence of a magnetic field they align themselves along the magnetic flux and this greatly restricts the movement of the fluid. As a result, its viscosity can rapidly change from very fluid to almost solid. Magneto-rheostatic fluids have been used in fast-acting clutches, shock absorbers and flow control systems.

Did you know?

Weight for weight, wood is twice as strong as steel when used as a structural material.

Did you know?

Shape memory alloys can be used as dental braces. The temperature in the mouth causes them to contract and exert a force on the teeth.

PLTS

Describing the properties of a given engineering material will help you develop your skills as a **reflective learner**.

A plug top

• **Electro-rheostatic fluids:** These behave in a similar way to magneto-rheostatic fluids, in that they become more viscous in the presence of a static electric field. The effect is again due to small particles in the liquid aligning themselves along the flux, which opposes movement of the fluid. Electro-rheostatic fluids are fast acting and can change from a liquid into a stiff gel and back again in milliseconds. They are being used in similar applications to magneto-rheostatic fluids.

BTEC ## Assessment activity 8.1

1. Describe the following properties, used to define the behaviour of engineering materials. **P1**

   - tensile strength
   - hardness
   - toughness/brittleness
   - malleability/ductility
   - electrical conductivity
   - ferromagnetism
   - corrosion resistance
   - solvent resistance
   - resistance to environmental degradation

2. Choose a ferrous metal and a non-ferrous metal from the following lists and describe their properties. **P2**

cast iron	aluminium	mild steel	copper
lead	brass	stainless steel	bronze
medium carbon steel		high carbon steel	

3. Choose an organic material, a thermoplastic, a thermosetting plastic and a smart material from the following lists and describe their properties. **P3**

oak	PVC	Bakelite	pine	GRP
nylon	Formica	plywood	PTFE	MDF
piezoelectric material	Perspex			
Kevlar composite	shape memory alloy			
electro-rheostatic fluid	magneto-rheostatic fluid			

4. In the plug top shown in the photograph, the pins are made from brass and the body is made from melamine. Choose one of these materials and explain in terms of its properties why it has been selected for this use. **M1**

8.2 Know how engineering materials are identified

As part of your work as an engineering technician you may need to interpret the material requirements given on engineering drawings, process plans and service schedules. These are often given in abbreviated form and it will be helpful if you know some of the more common ones. You may then have to draw raw material or components from stores. In the absence of a storekeeper you may have to do this yourself. You will have to pinpoint the location and update the stores records after you have drawn the material. Here again you may find that abbreviations are used.

8.2.1 Symbols and abbreviations

Figure 8.2 shows part of a typical title block on an engineering drawing, containing information on the material to be used.

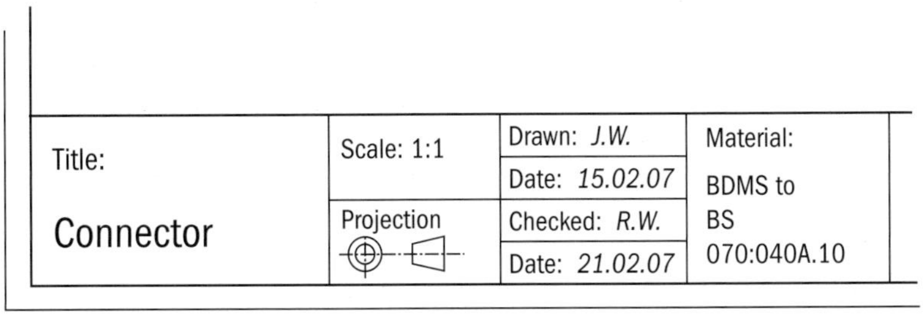

Title:	Scale: 1:1	Drawn: *J.W.*	Material:
		Date: *15.02.07*	BDMS to
Connector	Projection ⊕–◁	Checked: *R.W.*	BS
		Date: *21.02.07*	070:040A.10

Figure 8.2: A typical title block on an engineering drawing

The material specified in the title block is bright drawn mild steel, abbreviated to BDMS, together with its British Standard (BS) specification. Some other common abbreviations are shown in Table 8.4.

Abbreviation	Material
CI	Cast iron
SG Iron	Spheroidal graphite cast iron
MS	Mild steel
BDMS	Bright drawn mild steel
CRMS	Cold rolled mild steel
SS	Stainless steel
Alum	Aluminium
Dural	Duralumin
Phos Bronze	Phosphor-bronze

Table 8.4: Abbreviations for some common metals

A drawing or process plan may include additional information on the required surface finish, surface protection, heat treatment and surface hardness. For bar, sheet, pipe, wire and fastenings it may give some dimensional information. Some of these items are shown in Table 8.5.

Table 8.5: Material information

Abbreviation / symbol	Interpretation
ISO	International Organisation for Standardisation
BS	British Standard
BSI	British Standards Institution
BH	Brinell hardness number
VPN	Vickers pyramid hardness number
SWG	standard wire gauge
Ø50	50 mm diameter
MS, Hex Hd Bolt – M8 × 1.25 × 50	Mild steel hexagonal headed metric bolt 8 mm diameter, 1.25 mm pitch, 50 mm long

As in the title block shown in Figure 8.2, the British Standard or International Standard specification may be given for the material. This specifies the percentage of each of the ingredients of a material and its recommended uses. Surface hardness is tested by pressing an indenter into the surface of a material and using the dimensions of the resulting indentation to calculate a hardness number. In the Brinell test, a hardened steel ball is used and as the name suggests, a pyramid-shaped diamond is used in the Vickers pyramid test.

Standard wire gauge (SWG) is a means of classifying wire diameters and the thickness of sheet metal. The higher the SWG number, the smaller is the diameter or thickness. The most commonly used sizes lie in the range SWG16 (1.626 mm) to SWG42 (0.102 mm). The symbol Ø is used to indicate that a dimension is a diameter. For example, BDMS Ø50 would indicate bright drawn mild steel bar 50 mm in diameter. For pipe and tube sizes the outer and inner diameters might be given, or perhaps the outer diameter and wall thickness.

Metric bolts and nuts are always shown as indicated in Table 8.5. Additional information on drawings and service schedules might state, for example, that fastenings such as nuts, bolts, washers and machine screws must be cadmium plated or otherwise treated to give corrosion protection.

8.2.2 Forms of supply

Engineering metals begin their life as mineral ores. Plastics are derived from the by-products of oil distillation and from vegetable sources. Timber is obtained from forestry. These are the forms of supply that arrive for primary processing. Metals are smelted or otherwise extracted from their ores to produce ingots that go on for secondary processing. The raw materials for plastics are converted into powders, granules and resins, and timber is transported to the sawmills for cutting and seasoning before use.

Secondary processing may be applied at the same location, but often the materials are transported on to other manufacturers. One person's product is another person's raw material! Metal ingots are passed to rolling mills, where they are reheated and passed backwards and forwards between powerful rollers. Girders, railway lines and thick steel plate are produced in this way.

Once the hot rolled sections have been reduced in size, some of them are further processed by cold forming. This can be cold rolling to produce sheet metal and rectangular section bar, or cold drawing to produce round bar and thin-walled pipes such as the copper pipes used for plumbing. These are then stored ready for distribution to engineering companies and builders merchants. See Table 8.6.

Did you know?

At one time aluminium was regarded as a precious metal. In the court of Napoleon III of France in the 1840s, only the important guests were allowed aluminium cutlery. The remainder had to be satisfied with gold and silver.

Metals	Polymers	Timber
Ingots	Powders	Planks
Castings	Granules	Boards
Forgings	Resins	Composite sheets
Pressings	Sheet	Rods
Bars	Mouldings	
Sheet	Pipe and tube	
Plate	Film	
Pipe and tube		
Wire		
Rolled sections		
Extrusions		

Table 8.6: Forms of supply of materials

Some of the metal ingots and bars from the primary processing stage are delivered to specialist metal foundries for secondary forming into castings and forgings. Castings are made by pouring molten metal into sand moulds. There is also die-casting, in which the molten metal is injected into a shaped die. Aluminium alloys are particularly suited to this form of casting because of their low melting point.

In forging, roughly shaped billets of metal are heated and placed between shaped dies. Pressure is then applied by a drop-hammer or press, which squeezes the metal into shape. Castings and forgings are generally made to order and transported directly to engineering companies for final machining.

In a similar way the powders and resins for thermoplastic materials undergo secondary processing into sheets, moulded components, tubes, pipes and the different kinds of complicated hollow section that are used to make UPVC window frames and doors. This is done by a process called extrusion. The raw materials are heated until fluid and then forced through a specially shaped die. The process is rather like forcing toothpaste out of a tube. The emerging section is cooled, cut to length and stored ready to be transported to customers. Some malleable metals such as aluminium sections can also be formed by extrusion.

The products from secondary processing are supplied to companies that produce engineered products. Many smaller firms manufacture mouldings, fastenings and a variety of finished components, which they supply to larger firms. These might be motor and aircraft manufacturers, machine tool manufacturers, or many others who make domestic and industrial products. Firms such as these require a steady supply of raw materials from secondary processing in the forms shown in Table 8.6 and also a steady supply of finished components for assembly.

8.2.3 Identification coding

The British Standards Institution (BSI) issue codes that specify the constituents of the different kinds of metal and alloy in common use. In most cases they also specify the most appropriate uses and operating conditions, particularly for work at high temperatures and pressures. The codes have constantly been updated in consultation with industry. They are now gradually being replaced by the new European standards.

There has been slow uptake of the new European BS EN 10277 series of standards for steels, but no doubt they will become more widely used as time goes on. Many firms still use the BS 970 codes for steels (issued in 1991), or even the previous BS 970 codes (issued in 1955). It would have been a mammoth task to change all the old material codes on drawings and service schedules when the changes were made, so you need to be aware of the different coding systems.

Table 8.7 gives examples of the codes for some common grades of steel. As you can see, the systems are quite different. You can obtain the full material specification by referring to the latest standard. You will find that the same applies to the other metals and alloys, where the same material may be coded in different ways depending on the age of the drawing or schedule you are working to.

Material	BS EN 10277:1999	BS 970:1991	BS 970:1955
Mild steel	1.7021	210M15	EN 32M
Medium carbon steel	1.0511	080M40	EN 8
Tool steel	1.3505	534A99	EN 31
Free cutting steel	1.0715	230M07	EN 1A
High tensile steel	1.0407/1.1148	605M36T	EN16T

Table 8.7. Current and previous BS material codes for some common steels

Suppliers and manufacturers often have their own coding systems and stores location codes. Metal bars are often painted with a colour code on their ends so that they can be easily identified on storage racks. There is no universal system for this and you should familiarise yourself with the local system in use. You will usually find it displayed on a chart in a prominent position in the stores area. A typical system used by one major supplier is shown in Table 8.8, but this may not be the system in use at your place of work or training workshop.

Material	Colour code	
Mild steel	Red	●
Medium carbon steel	Yellow	○
High carbon steel	Purple/White	◑
Free cutting steel	Green	●
High tensile steel	White	○

Table 8.8: Typical colour codes for barstock

You may find that aluminium components are painted yellow. Again, this is not universal, but the result of spraying them with an etching primer that better enables the finishing coat of paint to adhere to the surface.

In the absence of a storekeeper you may be required to update the stores records when you draw material or components. This may be a manual entry on a record card, or a keyboard entry on the stores database. In either case it is important that you make the entry so that a low stock warning can be raised, manually or automatically, if the stock falls below a particular level.

Colour coded steel joists

BTEC **Assessment activity 8.2** P4 P5 M2 D1

1. Identify the abbreviations and symbols used on the assembly drawing shown in Figure 8.3. **P4**

Part No.	Description
1	CI Support
2	MS Shaft Ø25 mm
3	Phos. Bronze Brg.
4	4 off M6 × 1 × 25 Hex Hd. set screw & spring washer

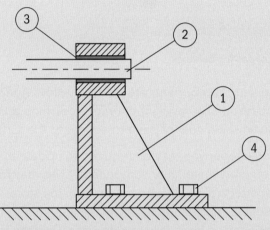

Figure 8.3: Bearing assembly

2. Identify the forms in which mild steel may be delivered to an engineering company. **P5**

3. Select an appropriate form of material supply for the manufacture of the phosphor-bronze bearing shown in Figure 8.3. **M2**

4. A sailing boat hull is to be moulded from glass-reinforced epoxy resin. Explain why this material has the required properties for the application. **D1**

SIMON WRIGHT
Materials engineer

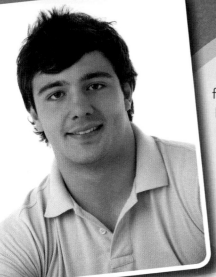

After leaving school at the age of 16, I wasn't sure what I wanted to do and toyed with the idea of a technical career in the army or air force. I achieved fairly good GCSE grades and decided that a full-time BTEC Level 3 National in Engineering would be a better option than studying for science A-levels. The course was broad based, with a high practical content and it was here that I first got interested in materials technology.

I completed the course with good grades in most of the units and was lucky to be accepted for technician training by a company producing precision sand and die castings. After an initial training period of one year my company sponsored me to study for a degree in materials engineering at university.

To begin with I found the course very demanding. Most of the other students had studied GCE A-level chemistry and I had to acquaint myself with this aspect of materials science. The facilities at university were excellent and I particularly enjoyed the practical metallography sessions.

On completing my degree I returned to my first company, where I stayed for almost three years, involved in the routine sampling of materials and the investigation of material faults reported by customers. I then moved to my current post with a company connected to the nuclear industry. My job involves the sampling and testing of materials and components that will be supplied to nuclear power stations. It also involves advising on the introduction of new or improved materials and making regular visits to our suppliers to discuss material specifications and requirements.

Materials technology is fast moving. New materials are constantly being developed to meet modern needs and new techniques are being developed for their manufacture and inspection. I get lot of job satisfaction and feel that I made the right choices at the start of my career.

Think about it!

- What parts of this chapter do you think would be particularly useful to an Engineering technician?
- What additional knowledge of the way in which materials are processed and tested do you think would be useful?
- Write a list and discuss with your peers.

Just checking

1. What is the difference between hardness and toughness in a material?

2. How can you distinguish between a ferrous metal and a non-ferrous metal?

3. What are the metals that are alloyed with copper to make brass and to make bronze?

4. How can you distinguish between a thermoplastic and a thermosetting plastic?

5. What kind of reinforcing materials do we use used with plastics?

6. Name and describe three kinds of wood composite.

7. What particular property does a piezoelectric material have and what kind of applications are these materials used for?

8. What do the abbreviations CI, BDMS and the symbol Ø stand for?

9. In what forms of supply are aluminium alloys available?

10. What do the letters ISO and BSI stand for?

edexcel ▦

Assignment tips

- Keep a tidy file of notes and a log of your laboratory or workshop activities that you can refer to when revising for an assessment.

- Recognise the areas where you are least confident and don't hesitate to seek help from your tutors.

- Read your assignment briefs carefully and make sure that you have addressed everything you have been asked to do.

- It is good practice to word-process your assignment work. It will develop your IT skills and make it easier for your tutor to assess your work and give you feedback.

Credit value: 10

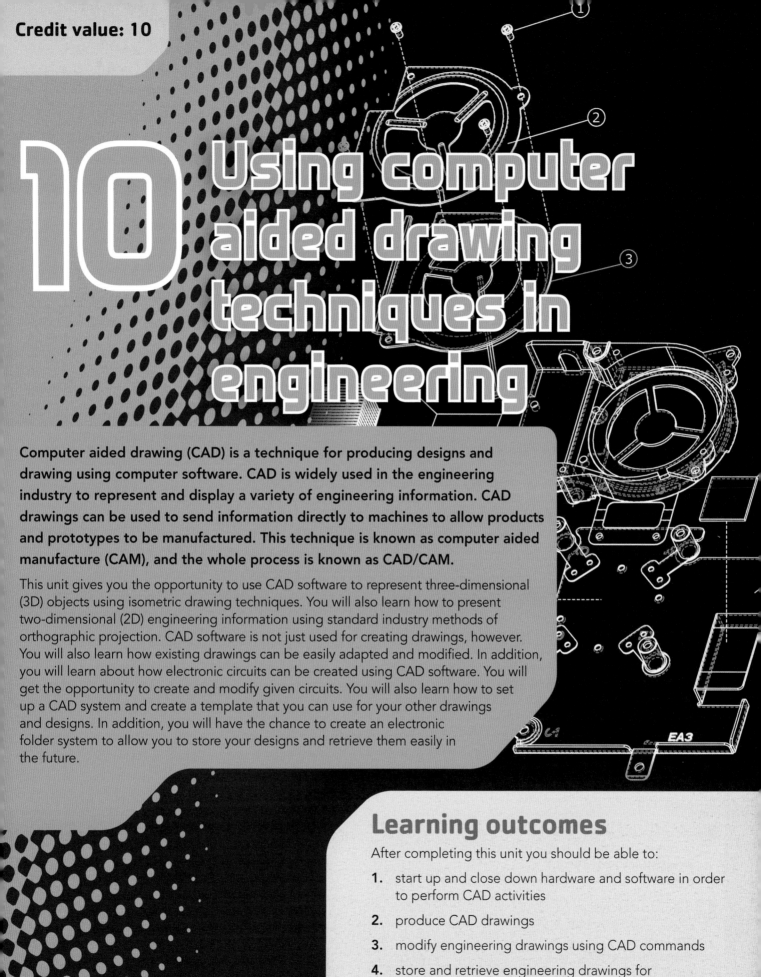

10 Using computer aided drawing techniques in engineering

Computer aided drawing (CAD) is a technique for producing designs and drawing using computer software. CAD is widely used in the engineering industry to represent and display a variety of engineering information. CAD drawings can be used to send information directly to machines to allow products and prototypes to be manufactured. This technique is known as computer aided manufacture (CAM), and the whole process is known as CAD/CAM.

This unit gives you the opportunity to use CAD software to represent three-dimensional (3D) objects using isometric drawing techniques. You will also learn how to present two-dimensional (2D) engineering information using standard industry methods of orthographic projection. CAD software is not just used for creating drawings, however. You will also learn how existing drawings can be easily adapted and modified. In addition, you will learn about how electronic circuits can be created using CAD software. You will get the opportunity to create and modify given circuits. You will also learn how to set up a CAD system and create a template that you can use for your other drawings and designs. In addition, you will have the chance to create an electronic folder system to allow you to store your designs and retrieve them easily in the future.

Learning outcomes

After completing this unit you should be able to:

1. start up and close down hardware and software in order to perform CAD activities

2. produce CAD drawings

3. modify engineering drawings using CAD commands

4. store and retrieve engineering drawings for printing/plotting.

Assessment and grading criteria

This table shows you what you must do in order to achieve a **pass**, **merit** or **distinction** and where you can find activities in this book to help you.

To achieve a **pass** grade the evidence must show that you are able to:	To achieve a **merit** grade the evidence must show that, in addition to the pass criteria, you are able to:	To achieve a **distinction** grade the evidence must show that, in addition to the pass and merit criteria, you are able to:
P1 start up a CAD system, produce and save a standard drawing template and close down CAD hardware and software in the approved manner **Assessment activity 10.1 page 160**	**M1** identify and describe the four methods used to overcome problems when starting up and closing down CAD hardware and software **Assessment activity 10.1 page 160**	**D1** justify the use of CAD for the production of a range of drawing types **Assessment activity 10.4 page 168**
P2 produce a CAD drawing using an orthographic projection method **Assessment activity 10.4 page 168**	**M2** describe the drawing commands used across the range of drawing types **Assessment activity 10.4 page 168**	**D2** demonstrate an ability to produce detailed and accurate drawings independently and within agreed timescales. **Assessment activity 10.2 page 162**
P3 produce a CAD drawing using an isometric projection method **Assessment activity 10.2 page 162**	**M3** describe the methods used to create relevant folder and file names and maintain directories to aid efficient recovery of data. **Assessment activity 10.7 page 178**	
P4 produce a circuit diagram using CAD **Assessment activity 10.3 page 165**		
P5 use CAD commands to modify a given orthographic and isometric drawing **Assessment activity 10.6 page 173**		
P6 use CAD commands to modify two different given circuit diagram types **Assessment activity 10.5 page 170**		
P7 set up an electronic folder for the storage and retrieval of information **Assessment activity 10.7 page 178**		
P8 store, retrieve and print/plot seven CAD-generated or modified drawings. **Assessment activity 10.7 page 178**		

How you will be assessed

This unit will be assessed by an internal assignment or assignments, which will be designed and marked by the staff at your centre. Assignments are designed to allow you to show your understanding of the unit outcomes. These relate to what you should be able to do after completing this unit.

Your assessment could be in the form of:

- presentations
- case studies
- electronic portfolios
- practical tasks.

Sarah, 16–year–old apprentice

This unit helped me see that you need to concentrate on what you want and that it takes practice, commitment and concentration to be successful.

I enjoyed using the software to produce a variety of different drawing types, and using the different drawing and modification tools. I learned about how to use the operating system to make files and folders, which made it easy for me to organise my work and find files for plotting. I realised how easy it is to take existing drawings and designs and use them to create new drawings. This is much quicker than starting from scratch and saves a lot of time that would otherwise be wasted.

There were lots of practical tasks and activities for this unit, so that made it more interesting for me. The part I enjoyed most was creating circuit diagrams and comparing these with real circuits. I liked using CAD software and found it easy to learn and operate.

Over to you

- What areas of this unit might you find challenging?
- Which section of the unit are you most looking forward to?
- What preparation can you do in readiness for the unit assessment(s)?

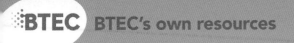

10.1 Be able to start up and close down hardware and software in order to perform CAD activities

Start up

Using template drawings

Think of times when you have looked at engineering drawings. What sorts of details are given on the drawing sheet, apart from the drawing itself?

Engineering drawings are produced using standard templates. Can you think of all the things that should be on a standard template? In addition, think of a CAD template. Would this be different from a paper template?

- What software settings should be preset?
- What information should be written on the template?
- What drawing sizes should be considered?
- What other features should a template have?

Produce a list and compare it with your colleagues' lists.

10.1.1 Start up

When you first switch on a computer it is important to check that all components of the system are operating effectively. You should check that all key elements are switched on and are operating correctly. The normal input devices such as a keyboard and mouse would normally be plugged in, although you should check this, of course.

If you are using a school or college network you should ensure that you have the correct details to allow you to log in and that you can access all the required network drives.

> ### Key term
>
> Specialist input/output devices – Devices attached to the computer such as a printer, digitiser, scanner or digital camera.

- **Specialist input devices** – Some CAD systems require a digitiser to input data. If this is required, you need to ensure that it is switched on, plugged in and configured correctly.

- **Specialist output devices** – Some CAD systems require a plotter or specialist printer, to allow large-scale and colour printouts to be produced. You should check that any such printing devices are switched on, plugged in and configured correctly.

Once the operating system has been activated, you will need to locate the CAD software you are using. This is often found using an icon on the desktop.

When you start the CAD software, you will probably be faced with an opening screen that will display a series of toolbars, drop-down menus and a drawing sheet (you can think of this as being like a piece of paper to draw on). In addition, you should see a cursor: this moves as you move your mouse or pointing device. You use it both to activate options from the menus and to use the draw and modify tools.

Snap/grid

When a grid is displayed, it allows you to position things. If you think about how graph paper makes it easier to plot lines, you can understand how useful the grid is. It is set at an even spacing: this is very useful if the snap function is also activated, as the mouse or pointing device you are using will go from dot to dot. It is usually sensible to set the snap and grid to be the same, and a 10 mm setting is a good starting point.

Drawing and erasing lines

With the grid and snap options selected, try to activate the option that allows you to draw lines. The cursor should move from dot to dot and allow you to draw quite complex shapes, although only in multiples of 10 mm.

Once you have drawn a few lines and produced some different shapes, it is useful to know how to remove them from the screen. By selecting the option that allows you to delete or erase items, you should be able to select the lines you have drawn and remove them. It is useful to practise this command, as you might be able to select all the lines you have drawn in one group to save time.

Absolute coordinates

Drawing using the snap and grid options is a good way to start and will give you some confidence in using the drawing tools, but to produce more accurate lines and shapes you will need to describe points in more detail. You use absolute coordinates in much the same way as you might plot a graph. There is an origin, which is normally in the bottom left-hand corner of the screen. This point is (0, 0), just as on a graph. Points are entered using the x–y system, so the point (100, 50) is 100 units in the x direction and 50 units in the y direction, both measured from the origin.

Individual CAD programs differ in the way that they operate. However, the command structure is fairly standard; you will need to get used to the idea of using commands to complete each operation and activating the next command to start the next operation.

This process often takes some getting used to and will take perseverance. It gets easier as you build up experience, and you will soon find you are producing drawings with confidence.

Did you know?

There are a wide variety of CAD packages available and each varies slightly in the way it operates. You will receive guidance and instruction from your tutor/assessor on how the software you are using will allow you to complete the activities in this chapter.

Did you know?

Undo is a very useful option, as it allows you to try something and experiment with the CAD tools and options. If it doesn't quite work out you can simply activate the Undo option and try again. Sometimes you can use Undo in the middle of a drawing operation to go back one step in the process.

Drawing in CAD is often a trial-and-error process, and you should not be afraid to try things out and learn from your mistakes. With a little perseverance you will soon be able to construct drawings with some confidence.

Key terms

Template – a standard layout that CAD operators use for producing drawings. It includes a border, a title block, and a standard logo for the company or organisation. It also features standard settings for such things as units, types of line, and colours.

Did you know?

Standard drawing sheet sizes are given A numbers:

A0 1189 mm × 841 mm

A1 841 mm × 594 mm

A2 594 mm × 420 mm

A3 420 mm × 297 mm

A4 297 mm × 210 mm

Worked example: Creating a template

This example shows you how to create a simple A4 **template** that you can use as a starting point for the other drawings you will be producing.

1. Activate the grid and snap and set both to 10 mm.

2. Use the draw/line option and when you are prompted for the starting point use the keyboard to input (10,10).

3. At the next point option input (287,10), then (287, 200), then (10, 200), and finally (10,10) again.

4. You should now have a rectangle that is equivalent to an A4 sheet of paper, but with a 10 mm border.

5. Using the zoom option, make sure you are looking at the bottom left-hand corner of the screen.

6. Active the draw/line option and use the snap and grid features to draw the title block, as shown in Figure 10.1.

7. Reset the snap/grid to 2 mm to add the text to the title block. Use a height of 2 mm for the text.

8. Save the template drawing, using the appropriate file format.

Name	Date
Title	Scale
	Projection
Drawing number	

Figure 10.1: Title block

Creating a template is one of the most useful activities a CAD user does. A template is used to begin a new drawing with the necessary layers and levels, types of text, dimension settings, types of line, and drawing layouts already prepared for the particular type of drawing. Different templates are used for large or small drawings, and for the various types of drawing and layout.

Layers and levels

CAD drawings use a system of levels to separate the various drawing elements. For example, you can have dimensions, text, guidelines, centrelines or hidden detail lines separated into different layers, which can be hidden from view when necessary to make the drawing clearer and therefore easier to understand.

Each industry uses its own layer standards to represent different things.

Using graphic images

Most CAD software packages allow you to insert images into your drawing. This is useful when you are creating a template, because you can place the logo for your company or organisation, for example, in the title block. This operation can be as simple as copying and pasting the logo into the drawing, but you may then have to resize and position it. Some software has special tools that allow you to place images with accuracy and precision.

Activity: Reading drawings

This activity requires you to use the template you created earlier. You will need to investigate the extra features of the CAD system you are using, and you should try to set up as many of the following options as you can:

- Set units to millimetres.
- Set layers and levels for different kinds of line, or for different coloured lines.
- Set grid and snap at 10 mm spacing.
- Add an extra space on the title block.
- Can you position the logo for *your* school or college in the extra space on the title block?

You may need to use the Help options in the software to help with this activity. When you have completed this exercise, see if you can save your drawing in a template or prototype format.

10.1.2 Close down

When you have finished your session, you should follow a standard routine when you close down the CAD system.

1. Save the drawing with the correct name, in the correct format, in the correct folder (use the Save as... option).

2. Close the drawing.

3. Exit from the CAD software (usually File → Exit).

4. Use the correct procedure to shut down the operating system. (You may need to log off first if you are using a school/college network.)

Assessment activity 10.1

Starting from when the computer system is first switched on, produce a flow chart showing all the stages of:

1. starting up a CAD system

2. opening a drawing template you have previously created

3. saving the template

4. closing down the software.

Your flow chart should include options to deal with the situation when a problem occurs with the hardware or software. **P1** **M1**

Grading tips

P1 You should feel confident that you can demonstrate how to start up and close down a CAD system if a tutor/assessor asks you to. You should also be able to produce a suitable drawing template. To achieve this grade you will need to draw a template and demonstrate starting up and closing down software and hardware.

M1 You should think of at least four things that could go wrong when starting up or closing down the CAD system and what you would if things went wrong. To achieve this grade you need to describe how you overcome problems encountered in starting up or closing down the system.

10.2 Be able to produce CAD drawings

Start up

CAD drawings

Think about a bicycle gear. This is an individual engineering component, but it is also part of a more complex assembly of components.

Workings in pairs, decide how you would draw the gear to allow it to be manufactured and checked for accuracy.

How many views would you need? Using a pencil, sketch out how you would present these views. What information do you think should be written on the drawing? How could a CAD drawing help you produce the component?

10.2.1 CAD drawings

CAD drawings are presented using the standard drawing conventions that have been used for many years. To represent objects in 3D we use a technique called isometric projection. To represent objects in detail we use 2D views, which are arranged in a conventional arrangement known as orthographic projection. And to allow us to understand how circuits work, we produce circuit diagrams.

Isometric projection

Isometric projection allows vertical lines to be drawn conventionally, with other lines drawn at 30° to the horizontal. What results is a 2D drawing that represents a 3D object. Conventionally, special paper is used to produce this type of drawing; however, CAD systems have settings and tools that allow you to quickly produce an isometric view.

Isometric projection allows you to draw lines at their correct lengths, and the CAD tools should allow you to direct the lines at 30° to the horizontal for lines to the right and left and straight up and down to show the height or depth of an object.

It is usual to be able to set parameters within the software to allow you to draw using an isometric grid and you should be able to set the system to have the cursor moving isometrically too.

 Did you know?

When constructing circles in isometric views you cannot use the normal circle option. This is because circles are represented by ellipses. CAD systems have a special tool for constructing isometric ellipses, although this is often an ellipse option.

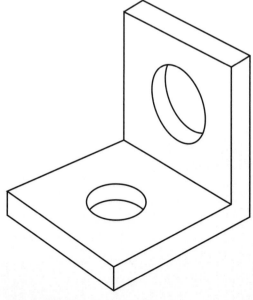

Figure 10.2: An isometric drawing

PLTS

When using a CAD package and being tasked with using tools and techniques you have not seen before, you will have to seek specific help and assistance from your tutor/assessor. This will help develop your skills as an **independent enquirer**.

By thinking creatively you should be able to see how you are going to produce a 2D image that will represent a 3D object. By using the tools available in the software you should be able to use trial and error to achieve your desired drawing. Both these will help develop your skills as a **creative thinker**.

BTEC **Assessment activity 10.2**

You should time yourself while you complete this activity.

Use the template you created in the worked example in section 10.1.1. Draw the bracket shown in isometric view in Figure 10.2. The sizes you should use are indicated in Figure 10.4. Each circle should be constructed using the correct technique and the short arcs that represent the rear of each hole should be visible. Save the drawing with the name Assessment 10.2.

Grading tips

P3 Isometric drawing requires the use of specific tools that allow lines to be drawn in three planes: left, right and top. To achieve this grade you will need to use the correct tools to produce an isometric drawing to the given sizes.

D2 To show an effective use of CAD you must be able to complete drawing exercises within a reasonable time. How long did it take you? Did you hurry and make any mistakes as a result? To achieve this grade you will need to produce detailed and accurate drawings by yourself in an agreed time.

Orthographic projection

When you need to show an object in detail it is usual to use a projection technique.

There are two commonly used projection systems: first angle projection and third angle projection. The projection system shows 2D views arranged and aligned in a particular manner. Normally there are three views, although sometimes extra views are used. Projection systems are always generated from the front view or front elevation. This is drawn to a given scale, and gives you your starting point.

Third angle projection:

- Starting with the front elevation, look at one side of this view and draw what you see on the same side.

- Starting with the front elevation, look from above this view and draw what you see above the front elevation.

First angle projection:

- Starting with the front elevation, look at one side of this view and draw what you see on the other side.

- Starting with the front elevation, look from above this view and draw what you see underneath the front elevation.

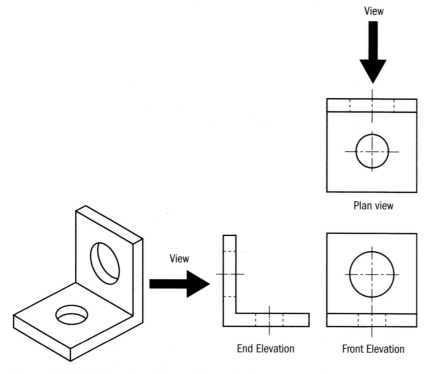

Figure 10.3: Third angle orthographic projection system

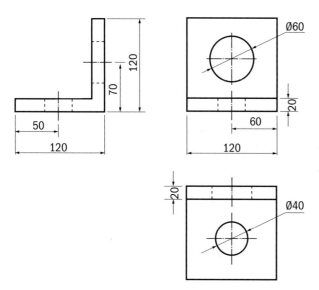

Figure 10.4: First angle orthographic projection system

As you can see from Figures 10.3 and 10.4 you can use different types of line to represent the various different details on a drawing. Table 10.1 shows the types of line that you can use, and when you would use them.

Table 10.1: The various types of line used in engineering drawings

Type of line	Appearance	Where used
────────	Continuous thick line	Visible outlines and edges
────────	Continuous thin line	Dimension, projection and leader lines, hatching, outlines of revolved sections, short centrelines, imaginary intersections
∼∼∼∼∼∼	Continuous thin wavy line	Limits of partial or interrupted views and sections, if the limit is not an axis
─/\/\─	Continuous thin straight with zigzags	Break lines
── ── ── ──	Dashed thin lines	Hidden outlines and edges
── ─ ── ─ ──	Chain thin lines	Centrelines, lines of symmetry, trajectories and loci, pitch lines and pitch circles
▬─ ─▬	Chain thin lines thick at the edges and changes of direction	Cutting planes
── ─ ── ─ ──	Chain thin double dashed line	Outlines and edges of adjacent parts, outlines and edges of alternative and extreme positions of movable parts, initial outlines prior to forming, bend lines on developed blanks or patterns

Circuit diagrams

Circuit diagrams are used to show the relationship between different parts in a circuit. They are drawn to indicate how things are connected together. Examples include electrical/electronic, hydraulic and pneumatic circuit diagrams.

CAD systems often have a library of parts that have been pre-drawn. This allows you to 'drag and drop' the symbols, and construct connecting lines between them.

It is often useful to use the grid function and position each component on the grid: this makes it easier to produce the linking lines that represent the pipework or wiring that connects the parts.

Table 10.2: Typical symbols used in circuits

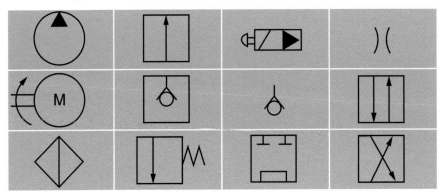

Did you know?

Circuit diagrams use symbols and formats that are specified by British Standards. For example:

BS 3939 Graphical symbols for electrical power, telecommunications and electronics diagrams. Guide for binary logic elements

BS 2917 Graphic symbols and circuit diagrams for fluid power systems and components. Specification for graphic symbols

Activity: Building blocks

When you are producing circuit diagrams, it helps if you have a library of parts to use, instead of having to draw everything from scratch. These symbols are sometimes called blocks. This exercise will allow you to build up a small library of components, which can be used to construct circuits later.

In your group, each person should draw at least one of the symbols shown in the table.

You will need to decide on sizes and you will need to save each symbol so that you can all access them later. This is where a network drive or shared file system helps.

Hint – you may find the grid function helpful.

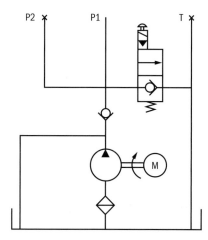

Figure 10.5: An example of a hydraulic circuit diagram

BTEC Assessment activity 10.3 P4

Draw the hydraulic circuit shown in Figure 10.5 using a library of pre-drawn parts. You should use the template you created in the worked example in 10.1.1.

Save the drawing as Assessment 10.3.

Grading tip

P4 You may need to share the symbols you have previously drawn with your colleagues, but the connecting lines, labels and other details will be your own work. To achieve this grade you will need to demonstrate that you can produce the circuit diagram using the correct CAD tools.

 PLTS

When using a CAD package and being tasked with using tools and techniques you have not seen before, you will have to seek specific help and assistance from your tutor/assessor. This will help develop your skills as an **independent enquirer**.

By thinking creatively you should be able to see how you are going to produce a circuit that will represent how the individual parts will be connected. By using the tools available in the software you should be able to use trial and error to achieve your desired drawing. This should help develop your skills as a **creative thinker**.

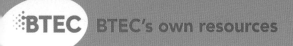

10.2.2 Drawing commands

The first thing you probably want to do when using a CAD system is draw some lines. There are systems that determine where you draw lines, how long they are and how accurate. To construct lines precisely you will need to have an understanding of the CAD coordinate system.

Absolute coordinates

We know that absolute coordinates are like map references. They are based on a datum position, which we call (0, 0). This understanding of the x and y coordinate system allows you to draw lines from or to precise points on the screen (x, y).

⇑ This is the positive y direction.   ⇓ This is the negative y direction.

⇒ This is the positive x direction.   ⇐ This is the negative x direction.

To draw using this system you need to know all the start and end points of every line. This is a slow method and it is easy to make mistakes.

Relative coordinates

Relative coordinates allow you to measure from your current point. So instead of having to work out the position all the time, you need only know the length of a line.

Polar coordinates

Instead of using the x–y system, polar coordinates require the length of the line and the angle it makes with the horizontal. Polar coordinates are also split into absolute and relative.

Figure 10.6 shows a triangle that has sides of 300, 400 and 500 units. Each of the three views represents the same triangle, but they have been drawn using these different methods of determining the coordinates.

Most standard CAD software packages contain a range of drawing commands. You can use the coordinate system to produce lines on the screen to exact sizes. This is the most basic and important function of the system, but there are a variety of other features that you should experiment with.

- Types of line – Different styles are available to represent different features. In Figure 10.3 different types of line are used to represent hidden detail and centrelines. It is often helpful to set different types of line for different layers.

- Circles – These are usually determined by specifying the centre point, and then the radius or diameter.

Did you know?

Attaching/snapping – CAD packages have a really useful tool that allows you to connect objects to each other. For example, you might have a circle that you want to position in the centre of another circle, or a line that needs to connect to the end of another line. This is useful not only for drawing, but also when moving or copying objects.

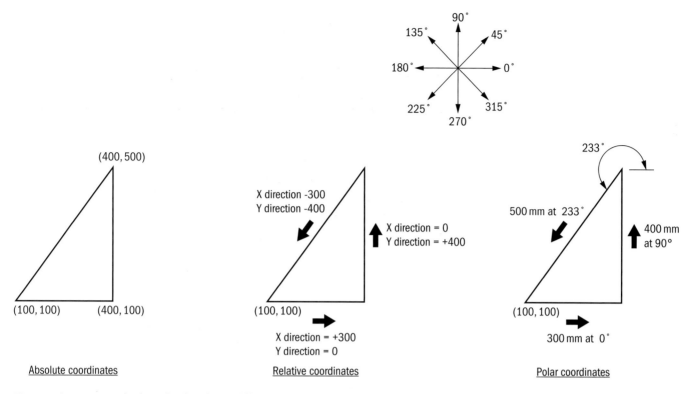

Figure 10.6: A triangle described in three different coordinate systems

- Arcs – Parts of circles can also be easily constructed using the radius and centre point or end point positions, angles etc.

- Rectangles – Rectangles, triangles, hexagons and other polygonal objects can be quickly constructed using the tools available.

- Text – CAD software features a text function that allows you to produce labels and add blocks of text.

- Hatching – Hatching is a technique that allows a regular pattern to be created. It is usually used to show where an object has been cut open to show the inside details.

- Zoom/pan – Zooming in and out allows you to view the whole of a drawing or particular details and panning allows you to move your view up/down or left/right to get a better idea of what you are looking at.

Did you know?

Most CAD packages allow you to add dimensions once you have completed a drawing. You can select features, and the software will automatically add the dimension lines, the leader lines and the correct dimension values. In this way you can very quickly add dimensions to a drawing. It is often useful to keep a layer specifically for dimensions, so that you can turn them off or hide them if necessary.

PLTS

When using a CAD package and being tasked with using tools and techniques you have not seen before, you will have to seek specific help and assistance from your tutor/assessor. This will help develop your skills as an **independent enquirer**.

By thinking creatively you should be able to see how you are going to produce a variety of 2D views aligned to represent the sizes of a 3D object. By using the tools available in the software you should be able to use trial and error to achieve your desired drawing. This should help develop your skills as a **creative thinker**.

Functional skills

Being able to create and modify different types of CAD drawing by yourself demonstrates the complex use of ICT systems.

Using the correct tools and techniques to create CAD drawings efficiently is a good way to demonstrate the use of ICT to present information.

BTEC Assessment activity 10.4

Using CAD software, draw the object shown in Figure 10.4 to full size. You should use the template you created in the worked example in section 10.1.1; you will also need to add the centrelines, hidden lines and dimensions. Draw the object in first angle projection. Save the completed drawing as Assessment 10.4.

As you complete the exercise, write down the individual tools and commands that you used to produce the different views.

When the drawing is complete, produce a short written statement to:

1. explain how you used each of the drawing commands

2. compare how you used each command with its use on the other drawings you have completed

3. compare the alternative, non-CAD methods of producing drawings with the use of CAD, and explain why you think CAD is the most useful in each case.

Grading tips

P2 You will need to make sure all the views line up properly. To help you do this, it might be useful to construct guidelines. To achieve this grade you will need to ensure that the drawing you produce has three views that are correctly positioned according to the conventions of orthographic projection.

M2 You will need to demonstrate how the use of commands is different when you are using commands to produce orthographic projections, isometric drawings, and circuit diagrams. It is also important to explain how these tools are activated, and how you used guidelines, connection tools and other techniques to speed up the drawing process. To achieve this grade you will need to demonstrate that you can describe the drawing command you use with different drawing types.

D1 This element requires you to justify the use of CAD. You should explain why you think it is the best method for producing drawings, and why it is better than other techniques. To achieve this grade you must justify the use of CAD for a range of drawing types, not just for one or two.

10.3 Be able to modify engineering drawings using CAD commands

Start up

Modifying drawings

Drawing using CAD can be considerably quicker than drawing by hand. One of the reasons for this is the range of tools that allow you to modify and manipulate drawings. Imagine you had to draw a series of fastening components such as nuts, bolts and washers. It would be very time consuming to draw each component individually, as it might be just the length or diameter that changed for each one.

Make a list of all the ways you could save time when you are drawing these fastener components.

10.3.1 Modifying drawings

Engineers often use two kinds of drawing.

Part drawings show all the necessary details to allow a part to be made. For example, you would produce part drawings for the individual parts of a bicycle, such as the gears or the seat post. A part drawing will be fully dimensioned, so that the part can be manufactured.

Assembly drawings are used to show how all the individual parts go together. Sometimes an exploded view drawing is used to show how an assembly fits together. Assembly drawings often have reference numbers (or balloon referencing); this allows the user to identify each individual part. The drawing will also contain a parts list. This is where the text function in CAD software helps, because you can usually draw up a table and edit the text very quickly.

If you have created a series of detail drawings in CAD, these can be copied to produce an assembly or exploded view.

Did you know?

Many engineering drawings have complex features that cannot be seen easily using conventional orthographic projection techniques. You can produce an extra view or partial view, sometimes shown at a larger scale, to allow clear details to be seen. CAD allows you to quickly copy and scale elements to produce these additional views.

Case study: Brian

You work for a company that manufactures alloy wheels for cars. You are investigating a new design for an alloy wheel with your colleague Brian. The alloy wheel you are considering is identical to a previous design, but has twelve spokes instead of eight. Brian is suggesting that he should start a new drawing. However, the current design is available as a CAD drawing that shows a three-view orthographic projection and an isometric view.

You suggest to Brian that he can use the existing CAD drawing to create the new wheel. Explain how he should go about this task.

Modifying circuit diagrams

When you construct a circuit diagram, you will find it easier to put the parts together if you are using a library of pre-drawn symbols or parts. A circuit can be constructed using the drag-and-drop technique. You will of course need to position the parts, add connecting lines, and possibly rotate or manipulate individual elements. Many circuits are very similar to existing circuits: it can be convenient to open an existing drawing, change or add the necessary components, and re-save the drawing with a different drawing number and description.

PLTS

When using a CAD package and being tasked with using tools and techniques you have not seen before, you will have to seek specific help and assistance from your tutor/assessor. This will help develop your skills as an **independent enquirer**.

By thinking creatively you should be able to see how you are going to edit an existing circuit to generate a new circuit and save yourself a significant amount of time and effort. By using the tools available in the software you should be able to use trial and error to achieve your desired drawing. This should help develop your skills as a **creative thinker**.

BTEC Assessment activity 10.5

Open the hydraulic circuit you produced for Assessment Activity 10.3.

Edit the drawing so that it is identical to the circuit shown in Figure 10.7.

When you have finished, save the drawing as Assessment 10.5.

Grading tip

P6 It is important to be able to take existing drawings and modify them in order to create new drawings. To achieve this grade you will need to show that you can modify two different circuit diagrams.

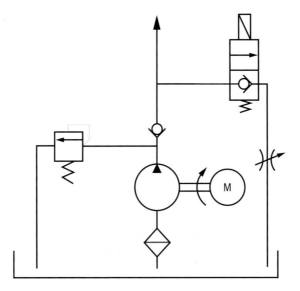

Figure 10.7: Modified hydraulic circuit diagram

10.3.2 Commands

There are many ways you can save time and increase efficiency when using a CAD system. 'Modify' tools allow drawings to be produced quickly and you should experiment with these techniques to see how you might be able to use them when you are constructing drawings or circuit diagrams. Figure 10.8 gives examples of the techniques you can use. You should try to reproduce some of the images shown here, but also try the other modify options listed and available to you.

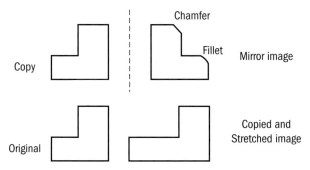

Figure 10.8: A few of the ways an existing image can be modified to produce a new one

Copy/duplicate

If you have constructed a drawn object and you need to duplicate it, you can do it quickly, easily and accurately with this command.

Array/pattern

If you have a feature that you need to produce multiple copies of, in either a rectangular or a circular pattern, you can quickly and easily reproduce it without having to draw or copy it over and over again.

Move

If you have drawn something in the wrong place, you can use the coordinate system to move it accurately and position it relative to other objects.

Rotate/revolve

If you draw an object that is shown in a particular position and then realise that it should be at a different angle, this command allows you to select it and then input the angle you need to rotate it by.

Erase

Inevitably you will make mistakes as you go along, or need to change something you have already drawn. The erase command allows you to remove unwanted items from the drawing. You can erase single objects, or select groups of objects to remove. If you erase something accidentally, the undo option will allow you to retrieve it.

Stretch

This is a useful option if you have a feature or group of objects that are too long or short. Activating this option allows you to stretch the objects to connect with other objects, or stretch them by a specific distance. You can often include the dimension lines as well and the dimension text will automatically adjust to the new measurement.

Trim

This is a useful option if you have been using guidelines or copy options and some lines overlap. It allows you to remove overlapping elements with reference to other objects on screen.

Scale

You can select an object and increase its size based around a **base point** you select. You can change the size of something relative to other objects, or by a given scale factor.

Chamfer and fillet

These functions allow you to change the appearance of a sharp corner. They are often confused with each other. The fillet option rounds off the sharp corner to a given radius, whereas chamfer cuts off the corner by a given angle or chamfer distance.

Change options

If you construct an element, and then realise that something is wrong, the properties of that object can be quickly and easily corrected. This saves you from having to redraw the object. For example, you can edit the radius of a circle, the colour of a line, the type of line or layer, or the type of hatch pattern. You can also match the properties of an object to those of another, similar object.

Key term

Base point – When you draw or modify objects, the base point is the key point of reference. For example, if you were drawing a circle, the centre of the circle would probably be your base point. Similarly, when you rotate an object you need a pivot or base point around which to rotate it.

BTEC Assessment activity 10.6 P5

This assessment is in two parts.

1. Isometric drawing

   Open the isometric drawing you created in Assessment Activity 10.2.

   Modify the drawing using appropriate techniques so that it matches the isometric view shown in Figure 10.9. You will notice that the holes in each face have been reversed. Save the drawing as Assessment 10.6a.

2. Orthographic projection

   Open the drawing you created in Assessment Activity 10.4.

   Modify the drawing so that it matches the drawing shown in Figure 10.10.

   Save the new drawing as Assessment 10.6b.

Grading tip

P5 Being able to edit existing entities quickly to create new drawings is a skill used by many CAD operators for speed and efficiency. To achieve this grade you will need to be able to modify a given orthographic and isometric drawing.

PLTS

In using a CAD package, and being tasked with using tools and techniques you have not seen before, you will have to seek specific help and assistance from your tutor/assessor. This will help develop your skills as an **independent enquirer**.

By thinking creatively you should be able to see how you are going to produce a 2D image that will represent a 3D object. By using the tools available in the software you should be able to use trial and error to achieve your desired drawing. This should help develop your skills as a **creative thinker.**

Functional skills

Being able to create and modify different types of CAD drawing by yourself demonstrates the complex use of ICT systems.

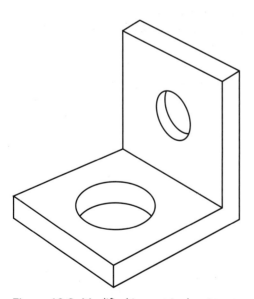

Figure 10.9: Modified isometric drawing

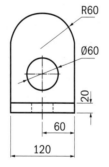

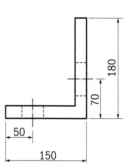

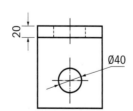

Figure 10.10: Modified orthographic drawing

10.4 Be able to store and retrieve engineering drawings for printing/plotting

Using hardware and software

Before computers were used to produce drawings and designs, engineering drawings were produced using drawing boards and conventional drawing techniques.

Write down five key factors that suggest that using computer aided drawing is a more efficient technique for producing drawings than conventional techniques. You should do this as an individual activity.

Compare your reasons with those listed by others in your group. Can you think of any reasons why conventional drawing techniques might be better than CAD techniques? Remember to consider the hardware and software.

10.4.1 Storage

The ability to store drawings safely and in the right place is essential for engineering organisations. Many organisations store thousands of drawings and may need to retrieve them at any time. Traditionally, a system of filing cabinets or special drawing cupboards was used to archive the hard copies, but electronic filing systems save a lot of space and are quicker and easier to access.

Storage media

Storage media are used for storing information or data. There are a variety of different methods for doing this: the main ones are listed in Table 10.3.

Portable storage media

Table 10.3: Storage media

Storage medium	Advantages	Disadvantages
Hard drive	Permanent fixed drive Large capacity	Only available on one computer
USB flash drive	Portable application; can be transferred between computers Compact Can have relatively large storage capacity Relatively cheap	Compact nature makes it easy to misplace
Network drive	Large storage capacity Accessible to all users on the computer network	Network connection needed to access files
Floppy disk	Portable Cheap	Small storage capacity Slow to operate High failure rate Few computers feature a floppy disk drive
CD/DVD	Portable Cheap Good for storing data that will not change (archiving)	Other devices have higher storage capacity Easy to misplace Reliability issues

Did you know?

Because of the value of engineering designs, many organisations store copies of their CAD drawings on portable storage media, such as portable hard drives, magnetic tapes or CD/DVDs. These copies are then stored in a fireproof safe. Should something happen to the computer network, the drawings can be retrieved. These copies are known as backup copies.

Key term

Computer network – a group of computers that are connected to a server, which is a central point for storing files and software. It allows controlled access to the network with passwords and login profiles for security, so that users can share files and folders.

The most-used method of storing engineering drawings that have been generated by CAD is the network drive. Files and folders are then accessible within the organisation to anyone on the **computer network**.

When a drawing is produced, it is given a unique reference number and is stored in a specific folder. To create folders it is often easiest to use the operating system. You should be able to use the file management system to find the area of the network drive or hard drive that you are using. Within this area will be an option to make/create a folder. Once you have created a folder, you can then open it and create sub-folders (folders inside folders) within it. Figure 10.11 shows how you could have folders for each subject you are studying, and sub-folders for the different types of drawing you have been producing.

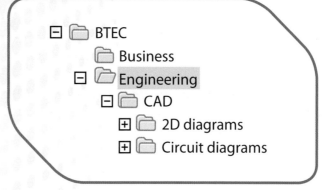

Figure 10.11: Folder structure

Activity: Creating folders

Locate the current location for your work.

Use the make/create folder option and create a folder called:

'Using computer aided drawing techniques in engineering'.

Open this folder and create four more folders:

'2D drawings'

'Circuit diagrams'

'Isometric drawing'

'Templates'

Using copy and paste, or other methods, move the CAD files you have already create into the appropriate folders.

Did you know?

Pen plotters move a pen across the surface of the paper or drawing film.

Inkjet printers work by spraying drops of ink onto the paper.

Laser printers operate by using a laser beam to produce separately charged areas of a drum. This allows ink to be fused to the paper in the charged areas.

10.4.2 Retrieval

Finding files to allow you to view, edit or print/plot them is straightforward if you have an organised file structure. You should be able to start the CAD software and use the option to open files that are normally found within the File menu. By browsing through the folders you have created, you should be able to locate any of the CAD drawings you have produced and open them to allow editing, printing and so on.

Plotting/printing

Producing hard copies of CAD drawings is often a more complex process than merely selecting the printer icon and then collecting the printout from the printer. Although you can use a normal A4 printer to reproduce your CAD work, it is often necessary to use a large-format printer or plotter.

When you have completed a CAD drawing you will probably want to produce a hard copy of it. Using the plot/print function will normally give you a series of options, which you should run through carefully, in sequence. Every CAD software package does things slightly differently, but the flow chart in Figure 10.12 shows a useful procedure/checklist to follow.

Did you know?

If you have lost a file, you can use the search facilities within the CAD package or the operating system to look for that type of file, or a specific named file.

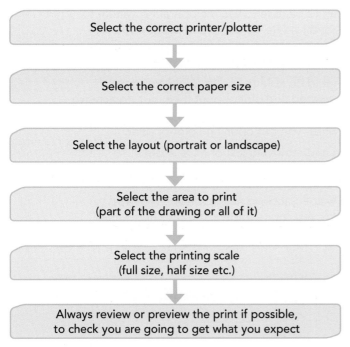

Figure 10.12: Checklist for printing out a CAD drawing

This activity is an observation. You should have your tutor/assessor or a colleague present and observing you.

Using a checklist, demonstrate that you can:

1. Create a folder with the name 'Activity 10.7'.

2. Save a CAD drawing of your choice in the folder with the name '10.7'.

3. Close the CAD drawing.

4. Retrieve the saved drawing from the 'Activity 10.7' folder.

5. Plot or print the drawing, using an appropriate scale to allow it to fit onto an A4 sheet of paper.

6. You should have created several drawings while you have been developing your CAD skills. Open a further six drawings and plot or print these drawings, making sure you have the correct scale to allow each drawing to fit on the sheets of paper you are using.

7. Close down the software and hardware in the correct way.

Describe the process, and explain how you can arrange all your CAD work and other files in a similar way.

Functional skills

Using a file/folder system helps you to develop your **ICT skills** in managing and finding information quickly and efficiently.

Plotting/printing drawings to the correct scale and at the correct size allows information to be easily viewed. You can also use the CAD system to transfer information or make objects easier to see.

Grading tips

P7 You should be able to use the operating system to create files and folders. To achieve this grade you will need to set up an electronic folder, save drawings into the folder and retrieve them afterwards.

P8 You should be able to find drawings you have previously created and plot or print them out. To achieve this grade you will need to have saved a total of seven drawings, which you can open and plot or print.

M3 Saving and retrieving just one file is fairly straightforward; however, you should be using this method to keep all of your computer work in an organised way. To achieve this grade you will need to describe how you create folders and save files within these folders. You will need to show how the file and folder names correspond, allowing you to find your work easily.

WorkSpace Graham Simpson

CAD technician

I work with a team of five other CAD technicians in the design office of an engineering company. My daily routine consists of:

- working on new designs
- making changes to engineering drawings
- participating in design meetings
- maintaining the file structure and checking that all files are saved correctly on the computer system
- testing the designs I have produced, in the workshop.

Typical day

A typical day for me involves arriving at the office for the morning meeting. We are given our allocation of design work for the day and usually spend time discussing any design issues or problems.

After working on the CAD system for about an hour, we always have a break to check that the plotters have enough paper and ink. It is important that we do not sit in front of the computers for too long without a break, as we have to consider health and safety.

At several times during the day we usually have to visit the workshop and observe prototype testing. This is where we make sure that the designs that we have developed as prototypes are performing as expected. Sometimes things don't go quite the way we expect and we have a look at the designs and try to work out what is going wrong. We usually spend a couple of hours during the day checking our designs with colleagues and having design meetings with one or two colleagues to share ideas, as some design problems can be difficult to solve.

The best thing about the job

I enjoy working on new designs and developing prototypes. Seeing products being made that I have been responsible for drawing is very rewarding.

Think about it!

1. Think of the activities that you have completed in this unit.
2. Which of these would be carried out on a day-to-day basis by a CAD technician?
3. What other skills might you need when working in a design office? Try to list all the things that might be useful for a CAD technician to know about.

Just checking

1. Describe the difference between a template file and a drawing file.

2. Explain how first and third angle projection are used in drawing layouts.

3. Explain why an isometric drawing would be used.

4. How can pre-drawn symbols be used in a circuit diagram?

5. What CAD commands are used to change the size of a drawn object?

6. What problems can occur when opening or closing CAD software?

7. Produce a list of all the drawing tools you might use when constructing CAD drawings.

8. Produce a list of all the modification options you might use when constructing CAD drawings.

9. Explain why CAD is used instead of traditional drawing board techniques.

Assignment tips

Much of the evidence that you will be collecting for this unit can be stored as an electronic portfolio. A portfolio is really a collection of files and folders that contain all the evidence you are going to present. You will need to organise your drawing files into folders and these can be stored on your computer network at school or college, or in the workplace.

- Keep backups – Make sure you keep a copy on a portable medium such as a USB flash drive.

- Use screen dumps – If you have created a file structure, or you want to show a feature of the software, you can take a snapshot of the image on screen. Some software packages allow you to select a certain part of the screen; alternatively you can press the Prt Scr (print screen) key on the keyboard. This will capture an image of the screen. You can then use the paste function to place the image into a basic word processing or graphics package.

- Save other files – Some of the activities require you to describe or justify the use of CAD features. You can produce word-processed files that can be stored with your CAD files and screen dumps.

- Keep it up to date – As you complete each activity, make sure you have a copy stored in the correct folder. When using complex software such as CAD it is worth saving your work at regular intervals: every 10 minutes is a good idea. Computer systems can crash and you would not want to lose all your hard work.

This unit links with Unit 2, Using and interpreting engineering information.

Some of the evidence you have collected in folders can be used to demonstrate how you:
- understand how to interpret drawings and related documentation

- use information from drawings and related documentation to carry out and check your own work output against a task specification.

You can also relate this evidence to personal, learning and thinking skills (PLTS). Examples of self-management can be demonstrated through portfolio building.

14 Selecting and using secondary machining techniques to remove material

Engineers design products to fulfil specific functions. To achieve this, they have to carefully control the size and shape of the product. Secondary machining techniques allow us to remove unwanted material in order to change the shape and form of a given workpiece. This is often by a combination of the four basic secondary machining techniques: drilling, milling, grinding and turning.

Before you carry out secondary machining operations, it is important that you position the workpiece correctly, and secure it firmly. During machining, you must follow the correct health and safety procedure. Each secondary machining operation has a different working principle: if you follow the correct procedure, you will ensure that risks are minimised.

This unit provides you with the opportunity to investigate how a range of secondary machining techniques are used. You will also look at how work-holding devices and tools are used in order to carry out the safe and effective machining of the workpiece. Using this knowledge you will have the opportunity to demonstrate the use of a secondary machining technique to make a workpiece. In this unit we shall also look at the health and safety aspects of secondary machining operations.

Learning outcomes

After completing this unit you should:

1. know how a range of secondary machining techniques is used

2. know how work-holding devices and tools are used

3. use a secondary machining technique safely and accurately to make a workpiece

4. know about aspects of health and safety relative to secondary machining techniques.

Assessment and grading criteria

This table shows you what you must do in order to achieve a **pass**, **merit** or **distinction** and where you can find activities in this book to help you.

To achieve a **pass** grade the evidence must show that you are able to:	To achieve a **merit** grade the evidence must show that, in addition to the pass criteria, you are able to:	To achieve a **distinction** grade the evidence must show that, in addition to the pass and merit criteria, you are able to:
P1 describe how three different secondary machining techniques are used **Assessment activity 14.1 page 188**	**M1** explain why it is important to carry out checks for accuracy of features on components during and after manufacture **Assessment activity 14.4 page 198**	**D1** justify the choice of a secondary machining technique for a given workpiece **Assessment activity 14.1 page 188** **Assessment activity 14.5 page 204**
P2 describe the appropriate use of three different work-holding devices for these different techniques **Assessment activity 14.2 page 193**	**M2** explain the importance of using the correct tooling and having machine parameters set correctly when machining a workpiece. **Assessment activity 14.5 page 204**	**D2** compare and contrast three secondary machining techniques for accuracy and safety of operation. **Assessment activity 14.5 page 204**
P3 describe the appropriate use of three different tools for these different techniques **Assessment activity 14.3 page 196**		
P4 monitor and adjust the machining parameters to machine a given workpiece correctly and safely and to produce features as defined by the workpiece **Assessment activity 14.4 page 198**		
P5 machine a given workpiece safely and carry out necessary checks for accuracy **Assessment activity 14.4 page 198**		
P6 describe methods of reducing risk for the secondary machining technique used. **Assessment activity 14.6 page 208**		

How you will be assessed

This unit will be assessed by an internal assignment or assignments, which will be designed and marked by the tutors at your centre. Assignments are designed to allow you to show your understanding of the unit outcomes. These relate to what you should be able to do after completing this unit.

Your assessment could be in the form of:

- presentations
- case studies
- practical tasks.

Claire, 16-year-old engineering apprentice

This unit helped me to see that you need to be determined to achieve your goals, and that it takes hard work, patience and practice to be successful.

I enjoyed investigating how things are made, and the different machining operations that are used to generate a finished product. It was good to see how a workpiece is positioned, and then do it for myself. I learned how the different tools are used, before machining starts. There were lots of practical tasks and activities for this unit, so that made it more interesting for me.

The bit I enjoyed most was setting up my work in the machine, using the tools we learned about, and then using the machine to make the piece I wanted, and being able to check it was the right size and shape. I liked participating in group activities and finding out about health and safety, and how we can protect ourselves from danger in the workshop.

Over to you

- Which areas of this unit might you find challenging?

- Which section of the unit are you most looking forward to?

- What preparation can you do in readiness for the unit assessment(s)?

14.1 Know how a range of secondary machining techniques is used

Start up

How is it made?

Think of an engineering component made from metal – something that might be part of an assembly, such as an engine or a bicycle. Write down four features that you think are important. Try to think how each one works, and how it fits together with other parts. For example, you might consider the surface, the shape, the function, or how the part is made.

Discuss your findings in small groups, and consider which machines might allow engineering components to be manufactured in order for them to perform correctly and allow them to fit together.

Did you know?

Machining operations often generate a lot of heat. This can damage cutting tools and affect the workpiece. This is why material is usually removed gradually, a little at a time. So you will usually need to make several cuts, sometimes with different tools, to achieve the final shape and surface finish you require.

Key term

Workpiece – the term we use to describe the part or component we are working on. Secondary machining operations are used to turn the workpiece into the finished part.

14.1.1 Secondary machining techniques

The manufacture of engineering products often requires the use of primary forming processes and secondary forming processes. Primary forming processes prepare raw materials for use. They include casting, forming, forging, extruding and moulding. Secondary forming processes usually follow primary forming processes, and involve material removal and change of shape. There are some very specific techniques, which are used in specialist workshops, but in this unit we are going to look at traditional secondary forming.

The secondary processes, and the machines that perform them, fall into four groups: turning, milling, drilling and grinding. These processes have something in common: they are all used in machining operations. So when you use these machines you are using secondary machining techniques.

Turning

Turning refers to the use of lathes to produce given shapes. The **workpiece** is usually cylindrical, and is located in a chuck and rotated. A cutting tool is applied to the surface to remove material.

Table 14.1: Turning operations

Operation	Purpose
Facing	Used to remove material from the end of a workpiece to leave a flat, square finish
Parting	Used when a cutting operation is required. Parting forms a deep groove, to allow a section to be removed from the whole workpiece. An example would be cutting off the head of a bolt
Turning	Used when a workpiece needs to have its outside diameter reduced
Drilling	Used to remove material from inside a workpiece. A drill bit is inserted into the tailstock and then moved into the workpiece
Screw cutting	Similar to turning, but the tool is applied to the workpiece at a specific angle to allow a thread to be created. Various tools are used, depending on the required thread

Types of lathe

- **Centre lathe:** this is used for individual turning operations, with a **tool change** between each operation.

- **Turret lathe:** this is similar in operation to a centre lathe, but instead of a tailstock it has a turret that can hold up to six cutting tools. The turret is usually mounted on a saddle, which slides directly on the bed of the lathe, and can be positioned ('indexed') by hand. This saves having to change tools, and means that the various steps of a machining operation can be carried out more quickly than on a centre lathe.

- **Capstan lathe:** The capstan lathe is similar in operation to a turret lathe, but is usually a smaller machine. The turret is mounted on a slide, which in turn is mounted on a saddle fixed to the bed of the lathe. The capstan lathe can automatically index between the tools for increased speed of operation.

- **Automatic lathe:** The automatic lathe is used in mass production. It may not look anything like a conventional lathe, and can have multiple tools, often controlled by computer. This allows the machine to be programmed to complete all cutting operations automatically.

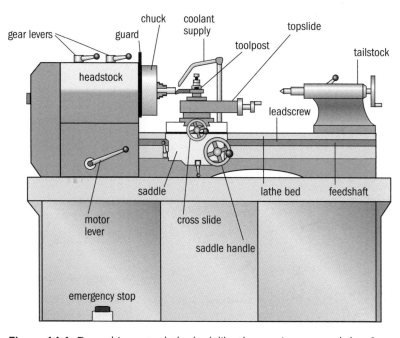

Figure 14.1: Does this centre lathe look like the one in your workshop?

Key term

Tool change – When machining on a lathe you will often need to perform a series of operations. You will need different tools for removing large amounts of material (roughing), for producing a precise surface finish (finishing), and for specialised operations such as facing and thread-cutting. This means that, for each different cutting operation, you will need to stop the machine, remove the guards, remove the tool, insert a different tool, align the cutting edge, secure the tool and replace the guards. This can all take a lot of time.

Key terms

Up-cut – a method of material removal where the cutter rotates in the opposite direction to the table feeding the work.

Down-cut – a method of material removal where the cutter rotates in the same direction as the table feeding the work.

Milling

Milling is a process of removing material in order to produce complex shapes. A milling machine consists of a rotating cutter with teeth that remove material. The workpiece is attached to the table of the machine with a vice and clamped tightly. The table or machine head can be moved up and down, sideways, forwards and backwards to bring the workpiece into contact with the cutter. The speed of the cutter is varied according to the type of shaping operation and the material being used.

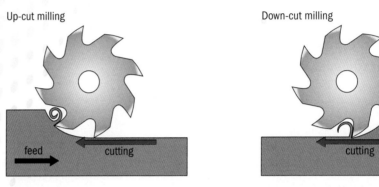

Up-cut milling   Down-cut milling

feed   cutting   cutting   feed

Figure 14.2: What is the basic difference between **up-cut** and **down-cut** milling?

Milling cutters come in a variety of sizes, shapes and types. The cutter can be changed quickly, depending upon the particular shaping operation and material being removed.

- **Vertical milling:** A vertical milling machine has the cutter mounted in a machine head that can move up and down. More complex vertical milling machines can work at different angles, and can produce chamfered edges on the workpiece.

- **Horizontal milling:** A horizontal milling machine is often used if less accuracy is required, or if a significant amount of material needs removing. The cutter is normally fixed in position, and the table is adjusted to line the workpiece up with the cutter.

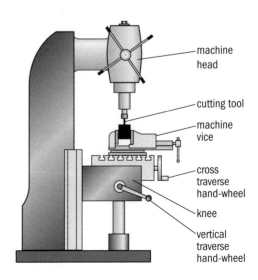

machine head

cutting tool

machine vice

cross traverse hand-wheel

knee

vertical traverse hand-wheel

Figure 14.3: The cutter can move up and down in a vertical milling machine

Drilling

Drilling is an operation to create circular holes in a workpiece. The cutter is called a drill bit, and it is positioned in a chuck. Drilling is a step-by-step process:

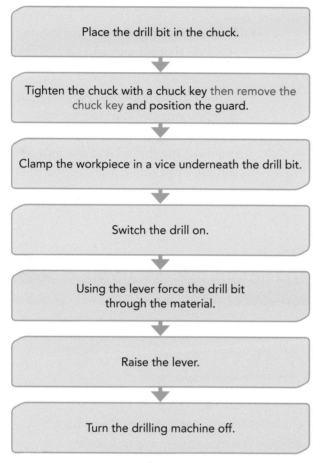

Place the drill bit in the chuck.

↓

Tighten the chuck with a chuck key then remove the chuck key and position the guard.

↓

Clamp the workpiece in a vice underneath the drill bit.

↓

Switch the drill on.

↓

Using the lever force the drill bit through the material.

↓

Raise the lever.

↓

Turn the drilling machine off.

Figure 14.4: The process of drilling

A drilling machine

Did you know?

Drills range from small-scale DIY drills to large-scale radial drilling machines. The most common drilling machines used in engineering are bench and pedestal drills. These are very similar in operation, but the bench drill is mounted on a workbench whereas the pedestal drill is free standing.

Grinding

Grinding is used for removing small amounts of material from the workpiece, and is often a finishing operation. An abrasive wheel is used as the cutting tool. Other machining operations cut the material, but grinding removes it by abrasion. The grinding wheel is powered by an electric motor, and rotates above the workpiece, which is fixed to the machine bed. Depending upon the machine, the workpiece can be moved towards the grinding wheel, or the grinding head can be moved across the fixed workpiece.

- **Surface grinding:** the workpiece moves from side to side as the grinding head is lowered until it makes contact with the workpiece.

- **Cylindrical grinding:** the workpiece is cylindrical and is held at each end, rather as it would be in a lathe. The grinding wheel contacts the workpiece to form a precise, circular polished surface.

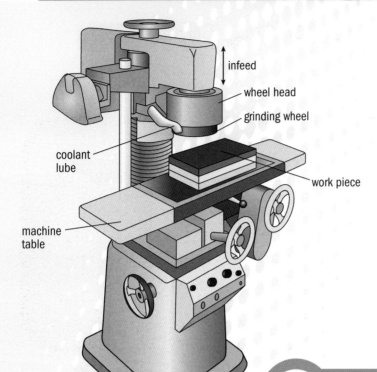

- **Centreless grinding:** this is a type of cylindrical grinding machine. Instead of supporting the workpiece at each end, it is held against the grinding wheel by a second, smaller wheel known as a regulating wheel. This rotates more slowly than the grinding wheel.

- **Profile grinding:** this is a technique in which specially shaped grinding wheels produce specific outlines or profiles.

- **Thread grinding:** this is performed using special grinding wheels fitted to a cylindrical grinding machine. The grinding wheels form the same profile as the thread and produce a high-quality surface finish with a high degree of accuracy.

Figure 14.5: Grinding machines are often used for finishing.

Functional skills

By evaluating three secondary machining techniques you will develop your **writing** skills.

BTEC **Assessment activity 14.1**

Select three different secondary machining techniques and describe how each one is used. Draw a flow chart to show:

1. how the workpiece is secured

2. how the tool approaches the workpiece

3. how the tool removes material from the workpiece

4. how the tool is removed from the workpiece.

In each box of the flow chart make sure you describe what is happening. **P1**

If you had a piece of round metal bar 150 mm long and 50 mm in diameter, which technique would you use to produce a hollow bore in the bar of diameter 25 mm? Explain why. **D1**

Grading tips

The way material is removed differs, depending on the technique you use. To achieve **P1** you will need to describe in your own words how this is done.

D1 To justify a particular secondary machining technique you will have to use evaluation skills. For example, you could use a lathe or a drill to perform drilling operations. Which of these do you think is the best option, and why? You will need to explain both ways of producing the hole and give reasons why one method is better than the other.

14.2 Know how work-holding devices and tools are used

Work holding

When you position work in a machine, you have to make sure it stays where you want it to stay. Secondary machining operations can generate large forces, and the workpiece has to be securely positioned to prevent it from moving.

Write down three methods that you know of for securing a workpiece. How can you use these methods for each of the secondary machining techniques: turning, milling, grinding and drilling?

Discuss your ideas in pairs, and consider similarities and differences between the various techniques.

14.2.1 Work-holding devices

Work-holding devices are used to position and secure the workpiece. You must ensure that you have the correct work-holding device, and that you use it correctly. If the workpiece moves during a secondary machining operation, then the result will be poor quality. Also, it can be very dangerous if the workpiece is not secure, because the tool might shatter or force it off the machine and cause an accident.

14.2.2 Work-holding devices used in turning

You can position and secure the workpiece on a lathe in a variety of different ways. The method you choose will depend on the size and shape of the workpiece, and on the turning operation you are undertaking.

The chuck

The chuck can grip the workpiece quickly and accurately. You secure the workpiece in the lathe by placing it in the jaws of the chuck and tightening them with a large key until they hold the workpiece securely. Chucks normally have three jaws, but they can have four or even six.

Types of chuck

- **Hard jaws** – these are usually made from hardened steel, and are often serrated to increase the grip. If they are used on soft material they can damage its surface when tightened.

Remember

It is important to remove the chuck key before starting the lathe.

- **Soft jaws** – these can typically be made from plastic, mild steel or aluminium, and can be machined so that they hold a particular workpiece.

- **Collet chucks** – a collet is a sleeve that normally has a cone-shaped outer surface and a cylindrical inner surface. The workpiece is placed inside the collet and centred. The collet has slots that allow it to contract and expand, so that as it is pushed or pulled into a second taper it contracts until it holds the workpiece securely.

- **Magnetic** – a magnetic chuck is used to hold iron-based materials that are too thin for, or may be damaged by, a conventional chuck. An accurately machined permanent magnet or electromagnet is used. Steel pole pieces are used to position the work accurately.

- **Pneumatic** – a pneumatic chuck is used to hold a non-magnetic workpiece that is thin or prone to damage. Air is pumped out of a space behind the workpiece causing a vacuum, so that atmospheric pressure holds the workpiece against the chuck.

- **Four-jaw chucks** – these are used for round, and unusually shaped workpieces. They are also better for machining square sections, as they can grip each side more easily than three-jaw chucks. The four jaws of the chuck can be adjusted independently.

- **Six-jaw chucks** – these are very useful if you have thin sections or hexagonal workpieces to grip.

- **Power chuck** – this replaces the chuck key. It normally has a pneumatic cylinder that allows air pressure to supply the power quickly to the jaws. This means you can clamp and release the workpiece more quickly.

Faceplates

A faceplate is a circular metal plate with slots machined into it. You can use it instead of a chuck; you use nuts and bolts to secure the workpiece to the plate. T-nuts are often used, as they can be locked into the slots. The faceplate fits onto the lathe spindle and is securely fixed by being screwed into place, held by a lock nut onto a taper or held by cam locks. The faceplate is not as convenient to use as the chuck, but it is useful for holding a range of difficult or awkward shapes.

Centres and driveplates

A lathe centre is a tool used to ensure that concentric work is produced. It is accurately machined at an angle of 60°. The workpiece has a hole drilled centrally in the axis, and the end of the centre is located in this hole. The end of the centre must be lubricated regularly to prevent it from friction-welding to the workpiece. A driveplate is similar to a small faceplate, and is used for driving work that is held between centres. It does not have to have an accurately machined surface, though, as it is used with a lathe dog.

Lathe dogs

A lathe dog is a device which is used to transmit the rotation of the driveplate to the work. It is clamped around the workpiece, and is then secured to the driveplate. It is sometimes known as a lathe carrier.

Fixed and travelling steadies

Steadies are used to support long workpieces, held in a chuck or between centres, which could easily deflect. They normally consist of three jaws with plastic or bronze rollers that can be adjusted to support the workpiece. A fixed steady is clamped to the bed of the lathe. A travelling steady is attached to a saddle that allows it to be moved along the workpiece during machining.

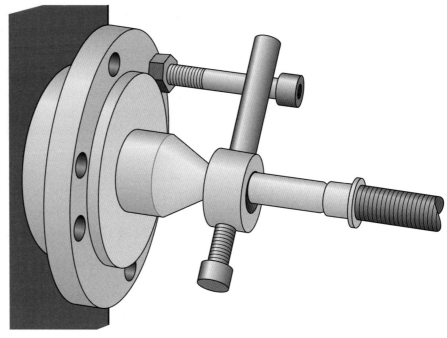

Figure 14.6: A lathe dog

14.2.3 Work-holding devices used in milling

When you use a milling machine, you can grip the workpiece in a variety of ways. Many of the methods are very similar to those already mentioned.

To fix the workpiece ready for a milling operation, you can:

* clamp it directly to the **machine table**.

* use a **pneumatic table**. This works on the same principle as the pneumatic chuck on a lathe.

* use a **machine vice**. This is very similar to a bench vice. It is clamped or fixed to the machine table, and has jaws that are brought together to hold the workpiece.

* use an **angle plate**. This is a plate with holes or slots that you attach the workpiece to. You then secure the angle plate to the bed of the milling machine. Angle plates can be adjustable so that you can tilt the workpiece if necessary.

* use a **magnetic table**. This works in a similar way to the magnetic chuck on a lathe.

* clamp the component in a **vee block**. This is either magnetic, or clamped in a vice. It is used primarily for milling circular workpieces.

 Did you know?

Fixtures can be used to hold the workpiece for a variety of secondary machining operations. A fixture is a device that is specially made to allow a particular workpiece to be effectively clamped or secured. Fixtures are quick and often simple to use, but they are very limited in the workpieces they can be used on.

- secure it in a **chuck**. This is used in the same way as the chuck on a lathe. It is normally the tool rather than the workpiece that is secured in the chuck of a milling machine.

- secure it in an **indexing head**. This allows movement through 90°. It looks like the driveplate and centre of a lathe, and is often used with a tailstock. The devices are clamped to the table of the milling machine, and the workpiece is positioned between them.

- clamp it in a **rotary table**. This is clamped in turn to the table of the milling machine. It allows you to machine at regular intervals around the work.

14.2.4 Work-holding devices used in drilling

Drilling operations are normally carried out with the workpiece clamped in a vice.

Other ways of holding the workpiece securely include:

- clamping directly to the machine table
- angle plate
- vee block and clamps
- fixtures.

14.2.5 Work-holding devices used in grinding

Grinding uses a variety of methods for clamping the work to the table. They include:

- chucks
- collets
- centres
- faceplates
- machine vices
- power chucks

- angle plates

- vee blocks and clamps (including magnetic vee blocks)

- fixtures

- work rests.

A work rest is a slotted base that can be tilted at a variety of angles. It supports the workpiece as it is pushed between the grinding wheels, aligning it with the centre line of the grinding wheels.

 Assessment activity 14.2

When you secure a workpiece you have to ensure that it will not move, slip or become dislodged. You should also make sure that the work-holding device is secure.

Describe how you would use a chuck, a machine vice and an angle plate to hold a workpiece in preparation for a secondary machining operation. Sketch each work-holding device, with an example of the type of workpiece you would expect to be secured. **P2**

Grading tip

P2 You should think about how different components need to be secured. For example, you would use vee blocks to support round components that could slip from a normal vice. To achieve this grade you will need to show you understand how each work-holding device is used.

14.2.6 Tools

A variety of different tools are used for the various different operations that you need to perform during secondary machining operations.

Tool materials

Tools come in a variety of materials, depending upon the material of the workpiece and the conditions the tool is being used in.

 Did you know?

Tools are what we use on a machine to remove the material from the workpiece. We fit different tools to a machine to carry out different operations.

Table 14.2 lists some examples of tool materials.

Material	Application
Solid high-speed steel	Used for a wide range of applications. It can withstand high temperatures without losing hardness. It is called 'high speed' because it can cut material much faster than high-carbon steels, and is resistant to wear.
Brazed tungsten carbide	Tungsten carbide is a very hard material that can cut faster than high-speed steel and produces a better surface finish. It is very expensive, and so only the cutting edge of the tool is made from tungsten carbide. This is attached to the steel body of the tool by brazing.
Composite wheels	Normal grinding wheels are constructed from particles of an abrasive material which are compressed and bonded together. Composite wheels use different materials mixed together with the abrasive to improve the surface finish of the workpiece or the performance of the wheel.

Table 14.2: Tool materials

Tools used in turning

You can use lathes for a variety of different operations, each of which needs a different tool. The diagram below shows the different types of operation and the tools required.

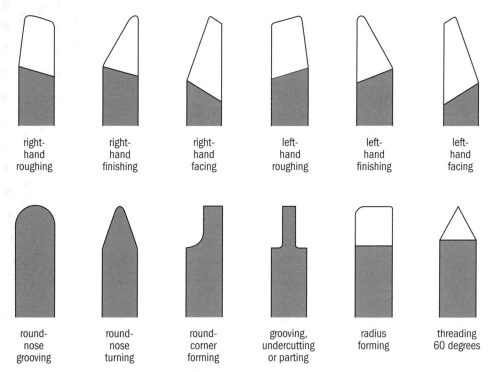

right-hand roughing · right-hand finishing · right-hand facing · left-hand roughing · left-hand finishing · left-hand facing

round-nose grooving · round-nose turning · round-corner forming · grooving, undercutting or parting · radius forming · threading 60 degrees

Figure 14.7: Each lathe operation needs a different tool

Tools used in milling

Milling machines are used for a wide range of different machining operations. They can produce parallel and perpendicular surfaces, or work at an angle. A wide range of milling cutters is available, for both horizontal milling and vertical milling.

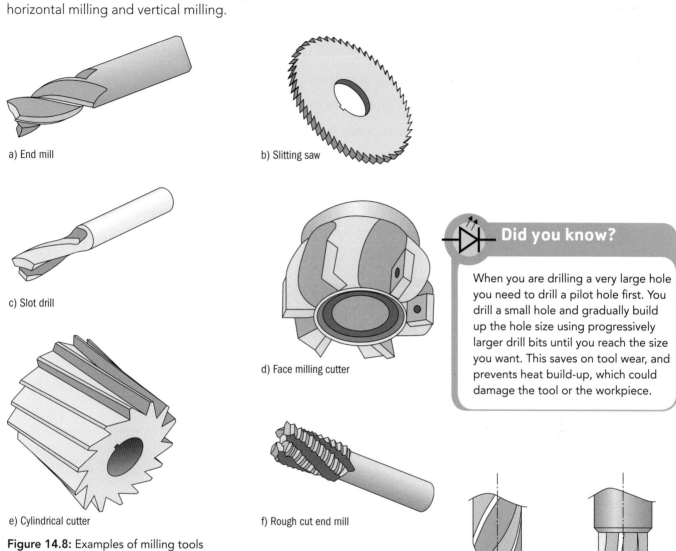

a) End mill

b) Slitting saw

c) Slot drill

d) Face milling cutter

e) Cylindrical cutter

f) Rough cut end mill

Figure 14.8: Examples of milling tools

Did you know?

When you are drilling a very large hole you need to drill a pilot hole first. You drill a small hole and gradually build up the hole size using progressively larger drill bits until you reach the size you want. This saves on tool wear, and prevents heat build-up, which could damage the tool or the workpiece.

Tools used in drilling

Drilling operations are sometimes carried out on a lathe. The rotating workpiece is moved to a stationary tool.

In general, though, in drilling operations the rotating tool (the drill bit) is brought into contact with the workpiece. We usually drill right through the workpiece, but sometimes we have to leave a 'blind hole' (one that does not go all the way through). We use a special drill to produce a flat-bottomed hole and use taps to cut threads in a hole. We can also use reamers for finishing holes, and special tools for countersinking, counterboring and spotfacing, as shown.

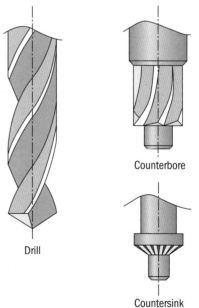

Drill

Counterbore

Countersink

Figure 14.9: Drilling tools: how many different drilling operations can you think of?

Tools used in grinding

A grinding wheel is made up of abrasive grains of material, held together by a bonding material.

Each grinding wheel has a specification that lists:

- the abrasive type
- the grit size (how big the grains of material are)
- the bond material
- the grade (how strong the bond is)
- the structure (how much bonding material is used).

A grinding wheel that has a very strong bond is known as a **hard wheel**. This is because it will wear slowly, but could generate too much heat in grinding. A wheel with a weak bond is known as a **soft wheel**: it will wear quickly, but will not overheat. Different shapes of grinding wheels are used for a variety of applications, and examples are shown in the photograph.

Grinding wheels come in many different sizes and shapes

Activity: Using different tools

For **one** of the secondary machining operations produce a sketch of three different tools and explain what they are used for.

To understand the use of a particular tool you may have to do some research. This could be by using the Internet and books in the library or learning resource area. However, it might be easier to speak to a tutor, assessor or technician who uses this type of tool regularly.

BTEC Assessment activity 14.3

Describe the tools you could use for creating threaded holes, milled slots and counterbores. Produce a sketch for each, and write a brief statement describing how each tool creates the desired shape.

Grading tip

You should be able to describe how each tool is used, which machine it is used on, and how it works. To achieve you will need to explain how different tools are used.

14.3 Be able to use a secondary machining technique safely and accurately to make a workpiece

Start up

Planning

You are given a piece of round steel bar 25 mm in diameter and 50 mm long. You are required to drill through it to form a hollow cylinder, and then produce an 8 mm metric thread (M8) all the way through. Write a step-by-step plan of how you would complete the task. You should explain the use of each tool and how it is fixed. Don't forget health and safety. You should also explain how the work is held in place and what procedures you would take to check that the finished piece is fit for purpose.

To complete this element you will need to demonstrate that you can use an appropriate secondary machining technique to produce a component from a given workpiece to the required degree of accuracy. Secondary machining techniques require a great deal of care. You should not use any machining techniques until you have had the correct training and health and safety instruction.

Your tutor will guide you on how to use at least one secondary machining technique. You should not work unsupervised until you have had some experience using all of the required tools and techniques. Even when you feel confident, you should operate machinery only when allowed by your tutor/assessor.

14.3.1 Machining parameters

Activity: Machining parameters

Show your tutor/supervisor that you can do each of the following:

1. Demonstrate the correct use of protective clothing and personal safety precautions.
2. Position a workpiece safely in its work-holding device and secure it.
3. Position and secure the tool in its tool holder.
4. Position the tool correctly in relation to the workpiece.
5. Adjust the position of coolant flow if necessary, and state the required flow rate
6. Demonstrate the safe and correct use of machine guards and safety mechanisms, including locating the emergency stops.

Case study: Maria

You have gained a work placement in a local engineering company. You have been asked to produce guidelines explaining how material removal is carried out in stages.

You have decided to produce a video to demonstrate the different techniques. Maria is a new apprentice, and you have decided to use her in the video, as she is good at explaining what is happening, and why she performs each machining operation.

Now answer the following questions:

1. What will you ask Maria to explain when she is drilling very large holes in metal plates?

2. How can Maria explain or demonstrate different techniques for metal removal in grinding, turning and milling?

3. What health and safety issues should you consider when making the video?

Did you know?

Secondary machining operations generate waste material. Drilling, milling and turning produce swarf. This is chipped pieces of material, which may be in continuous shavings or in small splinters. Swarf is a health and safety hazard – particularly sharp pieces of metal, which can cause cuts or more serious injuries. Swarf can also affect the performance of machining operations: it may get entangled in tooling, or be compressed by the tool and damage the surface of the workpiece. Care should be taken with the recycling of swarf.

PLTS

By monitoring and adjusting machine parameters to machine a given workpiece safely and correctly you will develop your skills as a **self manager**.

BTEC ### Assessment activity 14.4

Work in pairs. While your colleague carries out a secondary machining operation, you record the process, and check the accuracy of each step using appropriate instruments. Write down a step-by-step procedure of how the operation is carried out. When the process is complete, swap places: you perform the operation while your colleague records it. When you have both completed the process, discuss the importance of checking for accuracy – at each step, and at the end.

When you are writing down the process make sure you include:

* health and safety
* how the process was monitored
* how the machine, tooling and workpiece were adjusted
* what checks for accuracy were carried out. **P4** **P5** **M1**

Grading tips

Machining a workpiece correctly requires constant attention and adjustments. To achieve **P4** you will have to demonstrate you can carry out this process safely.

Once you have completed a secondary machining operation you must check that the workpiece is as required. To achieve **P5** you will need to machine a given workpiece accurately and check it for accuracy.

You should not wait until the final operation before you check to see whether a workpiece is correct. To achieve **M1** you will need to explain why frequent checks for accuracy are important.

14.3.2 Features of the workpiece

Before you use a machine you need to consider speeds and feeds. If you try to complete a machining operation too quickly you may cause damage, or produce a poor-quality finish. But in the workplace it is important to produce a high-quality finish in the shortest possible time, so you have to strike a balance between efficiency and quality.

Lathe cutting speeds

The speed of operation of a lathe depends on the finish required, on the depth of cut, on the type of tool being used, and on the physical properties of the tool and workpiece. You can calculate the speeds required and time taken using a simple formula.

 Worked example

Determine the spindle speed in **rev/min** when turning a 20 mm diameter bar at a **cutting speed** of 0.6 m/s.

The formula we use is:

$N = 1000S/\pi D$

where N = spindle speed (rev/min)

   S = cutting speed (m/min) = 0.6 × 60 = 36 m/min

   D = diameter (mm) = 20 mm

$N = (1000 \times 36)/(\pi \times 20)$

   = 573 rev/min (to the nearest rev/min)

 Worked example

Determine the time taken to turn a 30 mm diameter bar if it is 150 mm long, the cutting speed is 0.5 m/s, and the **feed rate** is 0.44 mm/rev.

$N = 1000S/\pi D$

where N = spindle speed (rev/min)

   S = cutting speed (m/min) = 0.5 × 60 = 30 m/min

   D = diameter (mm) = 30 mm

$N = (1000 \times 30)/(\pi \times 30)$

   = 318 rev/min (to the nearest rev/min)

Rate of feed can be converted into mm/min:

0.44 mm/rev × 318 rev/min = 140 mm/min (to the nearest mm)

To complete the machining of the 150 mm workpiece requires:

150 mm/(140mm/min) = 1.07 minutes × 60 = 64 seconds

 PLTS

By checking the machining parameters and making necessary adjustments to the way the workpiece is machined, you are constantly reviewing your progress and acting on the outcomes.

Checking the workpiece after each machining operation tells you whether you can proceed to the next step, or whether you will have to start again. In this way you are organising your time and resources, and prioritising your actions.

Key terms

Rev/min stands for revolutions (revs) per minute. In secondary machining operations it reflects:

- how quickly the workpiece is rotating in a lathe
- how quickly the drill bit rotates in a drilling operation
- how quickly the milling cutter rotates in a milling operation
- how quickly the grinding wheel rotates in a grinding operation

Feed rate refers to the distance the tool travels during one revolution of the workpiece, or the distance the workpiece travels during one revolution of the tool.

Cutting speed is the speed with which the workpiece moves relative to the tool.

Milling machine table feed

The calculations are different for a milling machine, because you are moving the table, not the tool. The table feed rate is adjusted according to the finish required, the amount of metal removed, the type of tool being used, and the physical properties of the tool and workpiece. You must also consider the power and capability of the machine. You can calculate the table feed rate with a formula similar to that used for turning.

Worked example

Determine the time taken to perform a cutting operation on a workpiece 200 mm long using a 100 mm diameter slab mill with six teeth and a feed per tooth of 0.04 mm. Assume the cutting speed is 0.7 m/s.

$N = 1000S/\pi D$

where N = spindle speed (rev/min)

   S = cutting speed (m/min) = $0.7 \times 60 = 42$ m/min

   D = diameter (mm) = 100 mm

$N = (1000 \times 42)/(\pi \times 100)$

   = 134 rev/min (to the nearest rev/min)

Rate of feed can be converted into mm/rev:

0.04 mm/tooth × 6 teeth = 0.24 mm/rev

The table feed rate = 0.24 mm/rev × 134 rev/min

= 32 mm/min (to the nearest mm)

The time taken to complete the cutting operation on the 200 mm workpiece is 200 mm/(32 mm/min)

= 6.25 minutes

= 6 minutes 15 seconds.

Drilling machine tool feed

In a pillar or bench drilling machine you can adjust the cutting speed by moving a drive belt onto different pulleys. You adjust the speed according to the finish required, the amount of metal removed, the type of tool being used and the physical properties of the tool and workpiece. You must also consider the power and capability of the machine. You can calculate the spindle speed and time taken using a simple formula similar to that used before.

Worked example

Determine the time taken for a 10 mm drill to cut through a mild steel plate 12 mm thick. The drill has been set at a cutting speed of 0.4 m/s and is being fed at a rate of 0.25 mm/rev.

$N = 1000S/\pi D$

where N = spindle speed (rev/min)

S = cutting speed (m/min) = 0.4 × 60 = 24 m/min

D = diameter (mm) = 10 mm

$N = (1000 × 24)/(\pi × 10)$

= 764 rev/min (to the nearest rev/min)

$t = 60P/NF$

where t = time (seconds)

P = depth of material cut (mm)

N = spindle speed (rev/min)

F = feed rate (mm/rev)

$t = (60 × 12)/(764 × 0.25) = 3.8$ seconds

14.3.3 Checks for accuracy

When you use secondary machining operations to produce components it is important to check that they are to the correct size. You have to ensure that they meet specification. Often this specification is an engineering drawing, on which the sizes and tolerances are given.

You will be given the opportunity to use a variety of measuring instruments, such as vernier callipers and micrometers. But before you use a measuring instrument you should first perform a quick visual check to ensure that the workpiece:

- has no minor imperfections
- has no sharp edges or burrs
- is not damaged, and does not contain swarf particles
- has no missing features
- has no scores or marks
- is square, and any threads are intact
- has no false cuts
- has steps in the section where they are supposed to be
- has holes in the correct position (do they look round?).

Functional skills

By researching the use of different secondary machining techniques you will develop your skills in **reading**.

BTEC **Assessment activity 14.5**

A particular machining process requires a workpiece to undergo four operations: drilling, milling, turning and grinding. Provide short written answers to the following questions:

1. Explain why you would check carefully that you have the right tooling, speeds and feed rates for each stage of the process. **M2**

2. You could perform the drilling operation on a lathe. Justify the use of a lathe instead of a drill. **D1**

3. Compare and contrast each process in terms of its accuracy and safety of operation. **D2**

Each machine can achieve different degrees of accuracy and surface finish. You may need to research this, or ask your tutor/assessor.

Grading tips

To produce a good-quality product you have to think carefully about which tooling to use, and the way the machine is set up, depending on whether the process is critical to quality or not. To achieve **M2** you will need to explain why different tools are used at different stages, and why you can set machine parameters to different values.

You should carefully consider the way different machines work, and explain the quality, time taken and accuracy of each process. It is then important to say which you think is best. To achieve **D1** you need to evaluate why you would choose a particular technique.

You should think about all four processes, and say how accurate and safe each one is. After you have written about each, you should try to evaluate all four in a short written statement. To achieve **D2** you will need to talk about different techniques, and say which ones are safer and which have the best accuracy.

14.4 Know about the aspects of health and safety relative to secondary machining techniques

Start up

Health and safety

Working in a group, try to think of all the risks to health and safety that could occur in an engineering workshop when performing secondary machining techniques. Have one member of the group write a list of all the risks. Are they the same for turning, milling, grinding and drilling, or does each technique have different hazards? Can you think of a way to prevent each hazard? Next to each risk write down your method of preventing the hazard occurring.

14.4.1 Health and safety

Engineering can be one of the most dangerous occupations you can undertake because of the nature of the machinery, equipment, tools and techniques used. It is essential to maintain safety for yourself, for other people working in the area and for visitors.

Health and safety is regulated by law, and it is the responsibility of all organisations to be aware of current legislation, of the particular regulations that apply to the industry they are involved in, of the materials they use and of the environment their employees will be working in.

The HASAWA is supported by ongoing legislation that is derived from European Directives. Here is a list of the key regulations:

- Health and Safety (First Aid) Regulations 1981
- Electricity at Work Regulations 1989
- Health and Safety Information for Employees Regulations 1989
- Health and Safety (Display Screen Equipment) Regulations 1992
- Manual Handling Operations Regulations 1992
- Personal Protective Equipment (PPE) at Work Regulations 1992 (as amended 2002)
- Workplace (Health, Safety and Welfare) Regulations 1992
- Reporting of Injuries, Diseases and Dangerous Occurrences Regulations (RIDDOR) 1995
- Confined Spaces Regulations 1997
- Lifting Operations and Lifting Equipment Regulations 1998

Did you know?

The Health and Safety at Work Act (HASAWA) was introduced in 1974. It is the single most important law in protecting your health and safety. It requires all employers, including schools and colleges, to ensure health and safety 'so far as is reasonably practicable'.

Key terms

Risk assessment – Risk assessments should be carried out by health and safety representatives or trained persons. There are five steps to a risk assessment:

1. Look for the hazards.
2. Decide who is likely to be harmed, and how.
3. Evaluate the risks, and decide whether sufficient precautions have been taken or whether further action is required.
4. Record your findings.
5. Review your assessment, and make revisions if necessary.

- Provision and Use of Work Equipment (PUWER) Regulations 1998
- Working Time Regulations 1998
- Management of Health and Safety at Work Regulations 1999
- Control of Substances Hazardous to Health (COSHH) Regulations 2002
- Control of Noise at Work Regulations 2005
- Supply of Machinery (Safety) (Amendment) Regulations 2005
- Employment Equality (Age) Regulations 2006.

Despite all these regulations, accidents can still happen. In order to try to prevent this happening, we carry out **risk assessments**.

Activity: Risk assessment

This is a classroom activity. Design a template that will allow you to complete the five steps of a risk assessment. Make sure you have a section for each of the five steps. Include sections for notes and actions to be taken.

Complete a risk assessment for the classroom.

Any hazard in the workplace can be harmful, but it may not always be a risk. For example, a can of petrol can be hazardous, but there is minimal risk if it is locked in a fireproof cupboard. However, if the petrol was poured into a bowl and was being used to clean components, close to a heat source, it *would* be considered a risk.

The use of coolant is a good example of managing risks. Coolants are non-toxic, but they can be an irritant to skin. Machine guards prevent coolant splashing onto the operator (i.e. you), and wearing overalls protects your clothes. However, you will be handling workpieces and tools that may have coolant on the surface, so you should also apply barrier cream to your hands. This forms a protective layer on your skin.

The Health and Safety Executive (HSE) publishes guides, templates and guidance on safe working practices. It also enforces the law and will investigate any serious accident that occurs in the workplace.

PLTS

By carrying out a risk assessment you are anticipating where a risk might occur, and by suggesting actions you are managing that risk.

14.4.2 Working safely

Before you carry out a secondary machining operation, a risk assessment should have been carried out for the work area. There are common features of secondary machining operations, and they are listed in the table below.

Table 14.3: Hazards in secondary machining

	Turning	Milling	Drilling	Grinding
Moving parts				
Machine guards				
Handling cutting fluids				
Insecure components				
Emergency stop				
Machine isolation				
Swarf disposal				
Tool/cutter breakage				
Airborne particles				
Appropriate protective clothing and equipment				

Each machine has its own particular hazards. For example, a grinding machine can generate sparks and grinding dust, so there should be sufficient ventilation and appropriate PPE should be worn, including eye protection.

Grinding metal using appropriate PPE

Remember

Working safely

- If you have long hair, always keep it tied back.
- Wear barrier cream to protect your hands.
- Do not lean on any machine, or place tools on it.
- Do not use any machine until you have received instruction and been given permission by your tutor/assessor.
- Do not lift heavy workpieces without assistance.
- Do not wear rings or bracelets while operating machinery.
- Ensure your overalls are in good condition, fit properly and are buttoned.
- Do not remove swarf with your hands.
- Always wear safety boots, goggles and other necessary personal protective equipment (PPE).

Did you know?

Machine guards are often interlocked. This means that the machine will not operate if you try to use it with the guard incorrectly fitted.

Activity: Working safely

For **one** of the secondary machining operations, complete the boxes in Table 14.3. Write a comment in each box to show you understand the importance of that feature, and the action you would take to maintain safe working at all times

Key term

CNC is short for Computer Numerical Control. CNC is used with complex machine tools to automate turning, milling, drilling and grinding operations. The operator doesn't have to move and fit the cutting tools; the computer controls all this automatically.

PLTS

By anticipating and managing risks when machining a workpiece you will develop your skills as a **reflective learner**.

BTEC

Assessment activity 14.6   P6

Choose one of the machines used for secondary machining techniques. Perform a risk assessment of the machine and the surrounding area.

Use a template for performing the risk assessment: don't forget the five steps. Consider how clean and tidy the work area is, as well as the machine. **P6**

Grading tip

This criterion requires you to describe ways of reducing risks for a secondary machining technique. When you do your risk assessment you must do two things:

1. identify the risks.

2. say what you would do to reduce the risk

To achieve **P6** you will need to describe how risk can be reduced.

WorkSpace **David Pearson**
Skilled Machinist

I work in a small workshop where we manufacture artificial joints for patients.

We are a small team and each of us can use a variety of the machines. The work is very individual, as we have to manufacture each piece to suit a particular patient. This means that every piece we make is a one-off.

Typical day

A typical day starts with the team getting prepared: putting our overalls on and putting barrier cream on our hands. We then gather together and the team leader hands out the job sheets for the day. Each day is different, depending upon the jobs we have been assigned to do. Although I usually work on the vertical milling machine, there can be days when I use the lathe or programme the **CNC** machines. We spend a lot of time measuring and checking, because we need to work to very close tolerances and this needs a high degree of accuracy; after all, the parts we make are going into a human body. The teamwork is great and we all try to help each other make the best-quality components possible.

The best thing about the job

I really enjoy working with unusual materials, which can be very challenging. The high precision and accuracy of the work are things that are very important to all of us. The feeling that the products I make are going to change somebody's life for the better is very rewarding. I've even had patients visit the workshop when they have had their operations in hospital. They like to see how the artificial hips, knees etc. are actually made.

Think about it!

1. What areas have you covered in this unit that provide you with the background knowledge and skills used by a Skilled Machinist?

2. What further skills might you need to develop? For example, you might consider here how you would ensure you make high quality components. Write a list and discuss in small groups.

Just checking

1. Describe how turning, milling, grinding and drilling remove material from a workpiece.
2. Describe three different ways in which you can secure a workpiece in place when preparing for a secondary machining operation.
3. Describe three different ways in which you can secure a tool in place when preparing for a secondary machining operation.
4. Describe how you would check the workpiece for accuracy during and after a machining operation.
5. State the five steps in completing a risk assessment.
6. Describe what can happen if the tooling or machine parameters are incorrect when you are undertake a secondary machining operation.

Assignment tips

When you are selecting the correct secondary machining technique for a given workpiece, there are several questions you need to ask yourself:

- What kind of work-holding device will allow you to carry out the activity?

- Is the required tooling specific to only one type of machine?

- How will you mount the tooling in the machine?

- Which technique will give you the best accuracy?

- Can the operation be carried out safely on all machine types?

- Will the workpiece require several operations on different machines?

You might find it useful to complete a planning sheet for the operation: this is a step by step guide to carrying out the operation. You can specify tooling, work-holding and machining operations, as well as special processes or treatments.

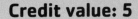

18 Engineering marking out

Engineering marking out is a technique used to ensure that, when components are manufactured, they are the correct size and shape and have the correct features. Marking out is a way of making marks on a workpiece or component to allow it to be accurately assembled or machined.

To mark out a workpiece accurately, you need to use a variety of tools and equipment. This can include the tools used for making marks on the material and the equipment used to guide the marking tools or support the workpiece. Marking out is a skill that requires accuracy and precision: this is why you should keep the tools and equipment that you use clean and store them safely when not in use.

Because marking out is an exercise that is often carried out in an engineering workshop, it is important to remember that safe working practices and health and safety considerations are of the utmost importance.

In this unit you will learn how to use marking out tools and equipment in a safe way to produce accurate lines and marks on a variety of different shapes and materials. You will also learn how to take appropriate care of the tools and equipment you use and how to plan your work carefully and methodically to ensure that all marking out activities meet the required specifications.

Learning outcomes

After completing this unit you should:

1. know about marking out methods and equipment for different applications

2. be able to mark out engineering workpieces to specification.

Assessment and grading criteria

This table shows you what you must do in order to achieve a **pass**, **merit** or **distinction** and where you can find activities in this book to help you.

To achieve a **pass** grade the evidence must show you are able to:	To achieve a **merit** grade the evidence must show that, in addition to the pass criteria, you are able to:	To achieve a **distinction** grade the evidence must show that, in addition to the pass and merit criteria, you are able to:
P1 select suitable measuring and marking out methods and equipment for three different applications **Assessment activity 18.2 page 223**	**M1** recommend corrective action for unsafe or defective marking out equipment **Assessment activity 18.1 page 222**	**D1** justify the choices of datum, work-holding equipment and measurement techniques used to mark out the three different applications. **Assessment activity 18.3 page 235**
P2 describe the measuring and marking out equipment used for the three different applications **Assessment activity 18.2 page 223**	**M2** carry out checks to ensure that the marked out components meet the requirements of the drawing or job description. **Assessment activity 18.3 page 235**	
P3 prepare a work plan for marking out each of the three different applications **Assessment activity 18.3 page 235**		
P4 mark out the three different applications to the prepared work plan **Assessment activity 18.3 page 235**		
P5 demonstrate safe working practices and good housekeeping. **Assessment activity 18.3 page 235**		

How you will be assessed

Your assessment could be in the form of:

- presentations
- written statements
- practical tasks.

Kasra, 16–year–old apprentice

This unit helped me to see that you need to be patient when learning new skills and techniques. I now realise that it takes practice, commitment and concentration to be successful.

I enjoyed using the different tools and equipment to mark out different components. It makes it much easier to perform secondary machining operations if the workpiece is clearly marked out. I learned about using the tools in different ways, depending on the application and the material.

I learned how to use the different clamps and supports to hold the workpiece steady and how to use different measuring techniques depending on the size and shape of the workpiece.

There were lots of practical tasks and activities for this unit, so that made it more interesting for me. The bit I enjoyed most was trying to work out how to hold difficult shapes securely so that they could be marked out. I also liked being able to use hand tools such as hammers and punches to show where drilling operations should take place.

Over to you

- What areas of this unit might you find challenging?
- Which section of the unit are you most looking forward to?
- What preparation can you do in readiness for the unit assessments?

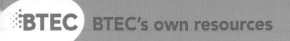

18.1 Know about marking out methods and equipment for different applications

From drawing to workpiece

Think of times when you have looked at engineering drawings. How does the information on the drawing get transferred to the material to allow accurate manufacture?

Engineering drawings are a form of template. They show the sizes and positions of different features on a workpiece. Imagine a drawing of a flat piece of material with a series of different-sized holes. Can you think of how you can ensure that all the holes are accurately positioned on the material?

- What tools would be needed to ensure the holes are accurately positioned?
- How would you make sure the drill bit is positioned correctly relative to the workpiece?
- Does the kind of material that you are using matter?
- What other issues should you consider?

Produce a list and compare it with those of your colleagues.

18.1.1 Measuring and marking out methods

If you are a carpenter or joiner working with wood, you need to use a pencil to mark the wood where you will be making cuts, drilling holes etc. In engineering it is not quite as easy, because you will be working with materials that are not so easy to make marks on. If you try to mark metals or plastics using a pencil, the marks can easily be rubbed off; the pencil will soon lose its sharpness and this means the lines will become too thick to be of use. Remember, you will usually be working to a high degree of accuracy.

Equipment required

If you were preparing to produce an engineering drawing, you would need a variety of drawing instruments and accessories: things such as propelling pencils, dividers, set squares, protractors, compasses, pencil sharpeners etc. When marking out you will need a toolkit of similar equipment to produce accurate lines and marks. This includes scribers, centre and dot punches, engineer's rule, callipers etc. This equipment is described in more detail later.

Key term

Workpiece – any component or piece of material that is to be machined or worked upon to create a final product.

Work–holding methods and devices

If you were preparing to produce an engineering drawing, you would need to secure the drawing sheet to a drawing board. You might use masking tape or special clips and you would make sure that the drawing board was clean and had a smooth surface or covering. This would ensure that the drawing sheet did not move as you created a neat, accurate drawing.

You need to use a similar technique when you are marking out. It is very important to secure the workpiece so that it does not move while you are marking out lines and centre marks. A drawing sheet is a two-dimensional (2D) object and is relatively easy to secure. An engineering workpiece can often be a complex three-dimensional (3D) shape and you will need to use a variety of devices, such as clamps, vee blocks and angle plates, to secure it.

Materials and consumables required

When you use a pencil or pen to draw on paper or drawing film, the lines are immediately clear and easy to see. But if you are marking engineering materials, such as steel or aluminium, it is sometimes difficult to see the marks you have made. To provide a clear background, dyes or lacquers are often applied to the surface before marking out. Alternatively, the lines can be made more prominent and easier to see by applying dye or powder. Other consumable materials are used to maintain the condition of tools and equipment. Oil or grease is often used and clean cloths or paper towels should be available to wipe clean tools and equipment.

Datum faces and reference points

When you draw a line on paper, you then measure from this line to the next line and so on. This original line is known as a **datum**. The same technique can be used when marking out. It is also often useful to measure from a particular face, edge or feature of a workpiece. This is also a datum and the equipment used for marking out can be set to reference this datum.

Key term

Datum – an edge, surface, centre, point, line or other feature that can be used as a reference to measure from.

18.1.2 Measuring and marking out tools and equipment

When you are working in an engineering environment it is always important to use the correct tools and equipment. This is equally true when you are marking out. To achieve high levels of precision and accuracy you cannot use the sort of drawing instruments that would be suitable for a classroom or drawing office. Instead, special tools and equipment are provided. Figure 18.1 shows some typical tools. You will get the opportunity to use tools like this when you perform marking out exercises.

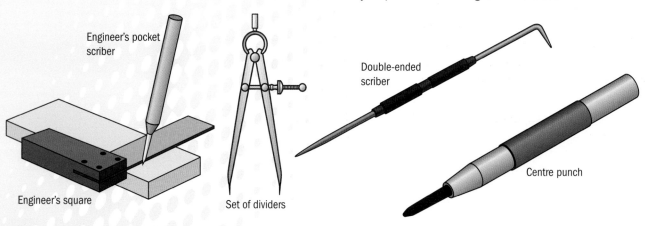

Engineer's pocket scriber

Double-ended scriber

Centre punch

Engineer's square

Set of dividers

Figure 18.1: Some typical marking out tools

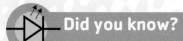
Did you know?

Just like a pencil, the point of a scriber should be kept sharp, to ensure that it produces accurate lines on the workpiece. An oilstone should be used regularly to sharpen it, because the heat generated if you use a grinding wheel is likely to soften the steel it is made of.

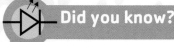
Did you know?

Just like the point of a scriber, the point of a punch should be kept sharp. It is possible to grind these tools, but this has to be done with care, to prevent softening of the material.

Scriber

When you draw on wood or paper, you use a pencil. However, when you need to 'draw' (i.e. make lines) on metal and on some plastics, you use a scriber. This is made from hardened steel and usually has a sharp point at each end, with one at 90° to the other. The scriber cuts a fine line into the surface of the material and this can be easily seen if the surface has been coated with ink, lacquer or whitewash.

Centre punch

Two kinds of centre punch are regularly used in marking out. A traditional centre punch is fairly heavy and has a shallow point. It is often used to locate the position of a twist drill. You may have tried to use a drill and found that the tip tends to 'wander' as you try to drill into the material. If you tap the end of a centre punch sharply with a hammer, the point will make an indentation in the material. You can then locate the drill in this indentation and it will not move when you apply power and force.

A dot punch is a smaller type of centre punch, which is used in a similar way to the centre punch, but has a sharper point. You can use it to create a mark for locating other instruments such as dividers or trammels, or you can use it to create a small mark, which can then be enlarged using the larger centre punch.

Engineer's square

An engineer's square is used to scribe a line at right angles to a datum edge or surface. It is constructed by securing a steel blade into a more substantial steel handle. The handle and the blade are both manufactured to a high standard, to ensure accuracy and precision in use. See Figure 18.1

Dividers

If you want to draw accurate circles on a piece of paper, you use a pair of compasses. To scribe a circle on engineering materials, you can use a pair of dividers. Dividers look like compasses and perform in a similar way. Each leg has a sharp point and the spacing between them can be accurately controlled using a screw thread and locking nut. The point of one leg of the dividers is located in a mark created by a dot punch; then, as the dividers are rotated, the point of the other leg scribes a circle or an arc. The legs of the dividers can be located in the graduations of an engineer's rule to allow accurate setting of the required radius.

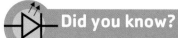

Did you know?

If very large arcs or circles are required, a trammel is used. This is a form of beam compass. It has an adjustable leg, which slides along the beam and a fixed leg, used for centring.

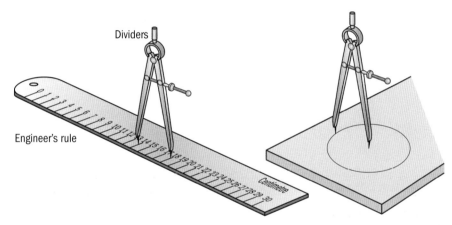

Figure 18.2: Dividers can be set with an engineer's rule and used to scribe circles or arcs

Odd–leg callipers

These are used to scribe a line parallel to a given edge or datum. They look like dividers and are adjusted in a similar way, but one of the divider legs is replaced by a bent calliper leg, which is located against the datum face for scribing. Odd-leg callipers are sometimes called hermaphrodite callipers or jenny callipers.

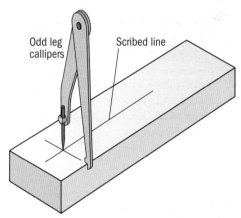

Figure 18.3: Using odd-leg callipers to mark against a datum

Engineer's rule

The sort of rule you might use in the classroom or home is frequently made from wood or plastic. An engineer's rule is not like this. It is a much more accurate tool, made from hardened and tempered steel and the graduations that mark the measurements are scribed accurately into the surface at millimetre intervals. The rule is accurately manufactured so as to have perfectly square edges and has a surface finish that is corrosion resistant. See Figure 18.2.

Did you know?

The end of an engineer's rule is precisely ground so that it can be used as a zero datum when measuring from a base surface or level.

Scribing block

A scribing block is sometimes called a surface gauge. It is used to mark out lines at a set height from a datum surface or edge. A scriber is clamped to the post of the scribing block so that, as the block is moved against the datum, a line is produced. The post can be adjusted to allow lines at an angle to be scribed. Pins are located in the base of the scribing block that can be lowered to allow them to locate against an edge and provide an alternative datum.

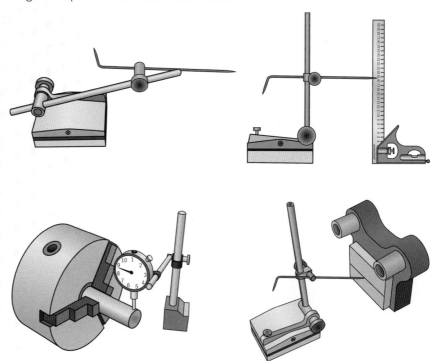

Figure 18.4: Different uses of the scribing block

Combination square

A combination square is similar to an engineer's rule, but it is more substantial, as it has interchangeable heads that can be clamped in place. The combination square is a used for a variety of laying out and marking out applications, depending upon the type of head that is fitted to it:

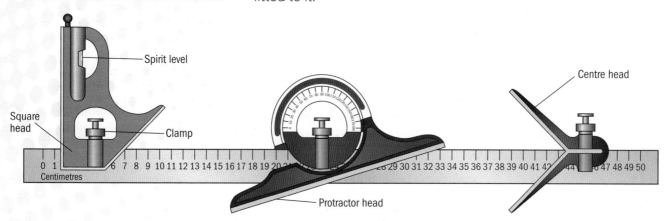

Figure 18.5: The different uses of a combination square

- Square head – used like an engineer's square to measure at 90° reliably; it can also be set at 45°.

- Centre head – used to find the centre of a circular section. It has two hard steel legs set at 90°. When it is placed on the circular section, a line can be scribed along the edge of the ruler, which passes through the centre of the bar. If this procedure is repeated in a different position, the point where the scribed lines intersect will be the centre of the circular section.

- Protractor head – this is graduated like a normal protractor and can be clamped to the combination square at any given angle. It allows lines at angles other than 90° or 45° to be accurately determined.

Vernier height gauge

A vernier height gauge works on the same principle as a vernier calliper. It is usually used to measure the height of components, but it can also be used to scribe lines parallel to a datum surface, because the pointer is sharpened to allow it to be used as a scriber.

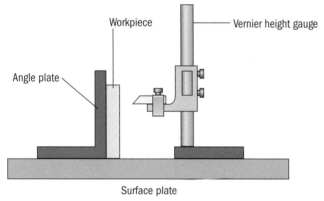

Figure 18.6: Using a vernier height gauge to measure height

Did you know?

You can use a scribing block or a vernier height gauge to scribe lines parallel to a base surface. However, the vernier height gauge has a built-in scale, whereas the scribing block has to be set to a separate steel ruler.

Slip gauges

Slip gauges are precision-ground steel blocks that have been machined to a very high tolerance using a process known as lapping. They are also heat-treated and are very accurate, with a very smooth surface finish. Slip gauges are used to aid accurate measuring, in inspection and in setting up accurate measuring equipment.

Slip gauges are supplied in a boxed set. The blocks are kept separate and when they are removed from their box and cleaned they can be 'wrung' together. This process requires you to bring together two gauges by rotating the face of one gauge relative to the other. Because of the quality of the surface finish the two gauges will effectively become one, as though they have been glued together. You can use this method to build up a stack of gauges to the size you desire, each one being carefully wrung to the next. This allows you to position marking out tools accurately relative to the workpiece.

Did you know?

If you leave slip gauges wrung together for too long they can cold-weld to each other and separating them will damage their precision faces. For this reason the blocks should be used quickly and always separated, lubricated and returned to their box when not in use.

Dial test indicator

A dial test indicator (DTI) is a measurement tool rather than a marking out tool. It records the displacement of a plunger that is very sensitive to movement. A rack and pinion mechanism in the DTI transforms this linear movement into angular movement, which is then amplified by a series of gears that turn a large pointer on the face of the DTI. A smaller circular display counts the revolutions of this large pointer. The main scale is typically subdivided into 0.01 mm increments, allowing tight tolerances to be checked.

Slip gauges are often used in conjunction with a DTI to check the size of a workpiece and the variation or tolerance. To separate the blocks you should slide one relative to the other.

Surface tables and plates

Most engineering workshops have an area put aside for marking out. The surface table is often the focal point for this activity. It is really a large flat table, which is usually manufactured from cast iron, as this is a smooth, self-lubricating surface, but glass, marble or granite can be used instead. It provides a smooth, level surface to act as a datum. It is very important to keep the surface table free of damage and it should be cleaned and covered when not in use.

A surface plate is a smaller version of the surface table. Although it is heavy, it can be moved onto a bench to allow marking out of smaller workpieces. The surface plate is also usually made from cast iron, with a smooth-machined working surface.

Angle plates

Work-holding devices are used to support the workpiece and the tools during marking out. It is important that all equipment and components being used are securely fixed and cannot be displaced or slip from their intended position.

Angle plates are usually constructed from cast iron and are machined accurately to form a right angle between the faces. They are slotted to allow T bolts to be used to secure work to them. Because they are heavy they can be relied on to hold work securely during marking out.

Vee blocks and clamps

Vee blocks are used to hold circular workpieces. They can be a simple vee shape, but the most common design incorporates slots down each side, which accommodate a special adjustable clamp to lock the workpiece in place. Vee blocks are manufactured in pairs, usually in cast iron or hardened steel. The pairs are matched and so should be carefully stored together.

Did you know?

Adjustable angle plates are available that can be used to locate a workpiece at an angle other than 90°. They can be set to the required angle using a built-in adjustment mechanism, but it is wise to use a vernier protractor to check that the angle is correct .

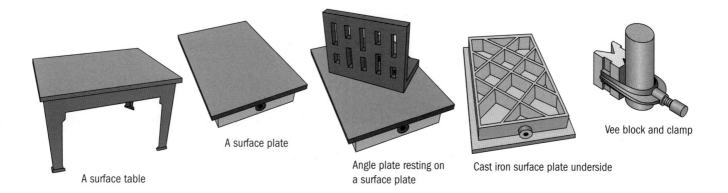

A surface table

A surface plate

Angle plate resting on a surface plate

Cast iron surface plate underside

Vee block and clamp

Figure 18.7: Work-holding devices: angle plates, surface plates and vee blocks

Calibration of measuring and marking out equipment

It is important to keep all tools and equipment used for marking out clean and damage free. If you use an engineer's rule for scraping or cleaning, it will become worn and damaged. You will not then be able to use it as a reliable measure.

By **calibrating** all marking out equipment regularly, you can ensure that the measurements you take are accurate and consistent.

Key term

Calibration – adjusting a measuring instrument, or checking it for accuracy, against a known standard.

Activity: Check it yourself

A good way to check whether measuring instruments are measuring accurately is to select a slip gauge, or a combination of slip gauges wrung together. You can measure the slip gauges using a variety of instruments, such as a micrometer, a vernier calliper, a vernier height gauge or a DTI. Do all the measurements you obtain match the value you expect? Any variation suggests that the particular instrument needs calibrating.

One question you should ask yourself, though, is whether the slip gauges can be relied upon.

Marking out media

To enhance the appearance of lines scribed on the surface of a workpiece, different types of dye or stain are used. The most commonly used is often called layout blue. A very thin layer is applied to metal; then, when a scriber is used, it scratches the blue off to leave a distinct mark. Do not confuse layout blue with engineer's blue, which is used to check mating parts.

When large castings or components are to be marked out it is usual to whitewash the workpiece first.

Assessment activity 18.1

You are supplied with a box containing a variety of tools used for marking out. Some are in very poor condition. Write an explanation of how you would check them for accuracy and describe what you would do if you found some were unsafe, or unfit for purpose.

Grading tip

M1 This activity requires you to consider the physical condition of marking out equipment as well as how well it performs. To achieve this grade you will need to recommend corrective action for unsafe or defective marking out equipment.

18.1.3 Applications

Having considered the different types of equipment available to help you when marking out materials, let us now look at the kinds of application you might encounter.

Square/rectangular

Many materials are supplied in long lengths called bar stock. For example, square and rectangular steel sections can be cut to the required length and clamped in position, ready for marking out. Care is needed if these sections are hollow, because excess clamping force could distort the material.

Sheet material

Steel plate is a typical sheet material, but this category of material could also include plastic sheeting. Sheet material can vary from substantial plate thickness to very thin materials. Marking out this type of material is more different than marking out on paper; however, the techniques you might use are similar.

Circular and cylindrical

Circular sections are commonly supplied in long lengths and cut to the required length. This material can vary from very large round sections to very small diameters. Once again, hollow sections should be handled and secured with care.

Irregular shapes

Unusually shaped objects can vary from simple applications, such as hexagonal-shaped bar, to unusual castings or forgings. Because of the one-off nature of some of these workpieces, a highly individual approach is often needed to ensure that the parts are correctly marked out. Adjustable jacks, clamps and wedges are often required to secure these types of component.

Assessment activity 18.2 (P1)(P2)

For the workpieces in Figures 18.8, 18.9 and 18.10:

1. List the measurement and marking out equipment you would use for each marking out activity.

2. Describe each piece of measuring and marking out equipment in your list and explain how it is used.

Grading tips

P1 To complete a marking out activity correctly you need to use the correct tools and equipment, depending on the type of workpiece you are marking out. You also need to be able to use the correct methods for marking out. To achieve this grade you will need to demonstrate that you can select the correct tools and equipment to complete the marking out activity and describe the marking out methods you would use.

P2 To use measuring and marking out tools and equipment effectively you need to understand how they are used and the settings, adjustments and features of each. To achieve this grade you will need to describe the measuring and marking out equipment you would use for each of the three activities.

Activity: Securing the workpiece

For each of the applications listed below suggest a way that you might secure the workpiece in preparation for marking out:

1. a solid aluminium section 40 mm square and 300 mm in length

2. a round steel section 20 mm in diameter and 100 mm long

3. a sheet of acrylic material 3 mm thick and 700 mm square.

PLTS

By working out the correct tools and equipment to use and deciding how to use them to complete a marking out exercise, you will show that you can identify questions to answer and problems to resolve. This will help develop your skills as an **independent enquirer**.

Mark out the profile on a piece of 10mm thick steel plate which measures 220mm × 70mm

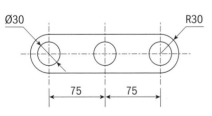

Figure 18.8: Marking out activity 1

Mark out 30mm diameter hole for the 60mm diameter bar supplied

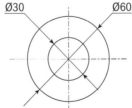

Figure 18.9: Marking out activity 2

The bracket is a casting with no holes. Mark out the holes to be drilled in the bracket.

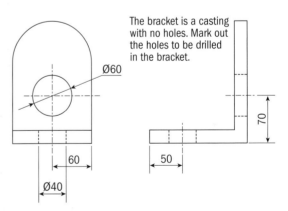

Figure 18.10: Marking out activity 3

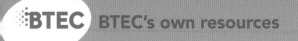

18.2 Be able to mark out engineering workpieces to specification

Activity: Planning

Take a sheet of A4 paper and, using appropriate techniques:

- Draw a circle, 100 mm in diameter, with the centre point in the centre of the sheet of paper.

- On the perimeter of the circle draw six equally spaced circles 40mm diameter in diameter.

- Check your drawing with that of a colleague or another student.

Now imagine that you had to complete this activity as a marking out exercise on a sheet of mild steel that is the same size as an A4 piece of paper. Write a sequence of operations to show the process you would undertake to complete this marking out exercise.

It is important to understand the methods and equipment used in marking out, but it is only when you carry out a practical marking out exercise that you begin to understand how difficult it can be. In this section you will learn how to prepare, set up and mark out workpieces to a planned sequence. You will also develop safe working practices and demonstrate the principles of good housekeeping to keep the work area and all tools and equipment clean and tidy.

18.2.1 Work plan

The process of marking out a workpiece can be very similar to producing an engineering drawing. Some of the equipment used in marking out is very similar to the instruments used in drawing. When you create a work plan you are deciding on the sequence of marking out operations and the materials and equipment you need. You can identify much of this information using an engineering drawing.

Use of digital vernier callipers on an engineering drawing

Reading engineering drawings or job instructions

To be able to mark out a workpiece using a drawing as a reference you need to understand and be able to read engineering drawings. There are many symbols and **abbreviations** commonly used in engineering drawing and you need to understand their meanings.

Unit 10, Using computer aided drawing techniques in engineering, gives guidance on the different types of line used in engineering drawings and how they are applied, as well as how different views are presented using orthographic projection.

Job instructions normally refer to a manufacturing plan for the workpiece. This details the sequence of secondary machining operations and can be used to help schedule the marking out process. However, you will often also need to refer to an engineering drawing to determine accurate positions.

Planning the sequence of marking out operations

To mark out a workpiece you should follow the sequence shown in Figure 18.11.

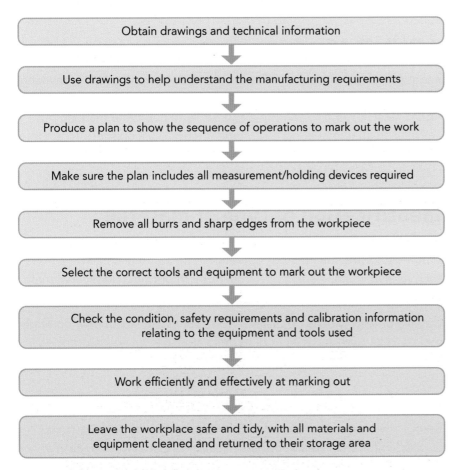

Obtain drawings and technical information

↓

Use drawings to help understand the manufacturing requirements

↓

Produce a plan to show the sequence of operations to mark out the work

↓

Make sure the plan includes all measurement/holding devices required

↓

Remove all burrs and sharp edges from the workpiece

↓

Select the correct tools and equipment to mark out the workpiece

↓

Check the condition, safety requirements and calibration information relating to the equipment and tools used

↓

Work efficiently and effectively at marking out

↓

Leave the workplace safe and tidy, with all materials and equipment cleaned and returned to their storage area

Figure 18.11: The sequence of marking out operations

Key terms

Typical engineering drawing abbreviations:

A/C – across corners

A/F – across flats

HEX HD – hexagon head

ASSY – assembly

CRS – centres

CL – centreline

CHAM – chamfer

CSK – countersunk

CBORE – counterbore

Ø – diameter (preceding a dimension)

R – radius (preceding a dimension, capital only)

PCD – pitch circle diameter

I/D – inside diameter

O/D – outside diameter

TYP – typical or typically

THK – thick

Clearly, much of this work is about preparation and planning. For example, when you are marking out a complex 3D part, you will need to decide which face to mark out first, how to clamp the work and how many times you will need to reorient the workpiece to complete the marking out process.

Identifying materials and equipment required

Part of the process of ensuring effective marking out is identifying the material you are marking out and the equipment you will need. You should have included this in your plan. Different materials could include steel, aluminium, cast iron or plastics.

You might need to use different techniques, depending upon the material of the workpiece:

- Would you use a heavy clamp to secure thin plastic sections?

- A heavy cast iron component might need a crane or lifting device to position it ready for marking out, but will a surface plate support it?

- Would you need the same force to centre-punch aluminium, plastic and steel?

- What about using a hammer action on brittle materials such as cast iron?

18.2.2 Marking out

It is important to understand the tools and equipment used for marking out, but it is only with practice that you can become proficient at it. Your tutor/assessor will give you the opportunity to complete marking out activities in an engineering environment such as a workshop. To achieve the required standard it is important to take your time, use a step-by-step approach and follow your work plan.

Identification of type of material

The first step in marking out is to identify the material you are using and prepare it for the marking out operation. The material could be a metal, such as aluminium, steel or cast iron, but it might be a plastic material.

These will need different methods. For example, you can use wax crayons or felt-tip pens on plastics. You should use layout blue on bright metals, but rough surfaces or black mild steel can have chalk rubbed into the surface.

Checking for visual defects

Before a workpiece can be marked out or machined, it is important to check it for any damage or defects, such as porosity or holes in a casting. Sheets of material should be checked for damage such as

Did you know?

Layout ink is often called marking blue, because it is often blue in colour.

cracks, or an uneven surface. A visual inspection is confined to looking for defects that can be seen by the naked eye; it does not identify problems that may be concealed beneath the surface of the material.

Cleaning

All components should be cleaned before marking out. It is important to remove:

- protective coatings

- rust

- grease

- dust.

This is because any coating on the materials will prevent the layout ink from sticking to the surface and will make marked out features difficult to see.

Deburring and sharp edges

Sharp edges can often be left behind after cutting operations, particularly when cutting metal: any sharp edge or splinter can cause deep cuts and must be removed. Similarly, a burr is a raised edge or splinter of material that has not been removed: this usually happens after a secondary machining operation such as grinding or drilling of a metal workpiece.

Coating

Once you have cleaned the workpiece, you should coat it with whitewash, layout ink or another suitable medium so that the marking out lines and features are clearly visible.

Setting and positioning the workpiece

Once the workpiece has been suitably prepared, you can secure it ready for marking out. First you should ensure you have the correct datum surface to work from. This is a perfectly flat surface for the workpiece and the tools and equipment. It would normally be a surface plate or surface table, although you also might need to support the workpiece in Vee blocks or using an angle plate. You should use the following equipment to ensure you have set up and positioned your workpiece securely.

- Use an engineer's square to ensure that the edges of the workpiece are square and that it is square to an angle plate or fixture.

- Use clamps and fixing bolts to secure the workpiece to angle plates or Vee blocks.

- Use a DTI to set up the workpiece and align work-holding devices.

- Use slip gauges with DTIs to check measurements and heights.

Key term

Deburring – a finishing operation to remove burrs, jagged and raised edges or surface imperfections; usually performed with careful use of a file.

- When you clamp a workpiece in place it can be useful to use packing pieces to support the clamp and help to wedge the workpiece to allow it to be supported or secured. Alternatively you can use them to increase the height or spacing between a workpiece and the marking out tools.

- Use a jack to support any work that is in danger of tipping over, or to provide additional support.

Marking out to a planned sequence of operations

It is important to follow your work plan; after you complete each step, make sure you check that you have marked out everything correctly. This can save you time and prevent incorrect machining of a workpiece, costing time and money.

Datum points

We have previously discussed the use of a datum as a point, edge, line or other element that is used as a reference for measuring from. A datum might be:

- Point datum – a single point, often the centre of a circle, from which other points or features are measured.

- Line datum – a line often scribed across a workpiece, from which other features or datums can be measured.

- Edge datum – this can be a surface datum and is frequently required when marking out unusual, irregular or 3D components. The edge can often be the surface of the marking out table or the edge of an angle plate.

 Did you know?

Engineers often use the expression 'Measure twice, cut once'. This means you should always double- check your measurements before you scribe or mark a workpiece.

Worked example: Marking out sheet material

This worked exaple sets out the procedure you need to use to mark out the profile shown in Figure 18.12.

1. Use a rule to check that the material is large enough to allow the component to be produced.

2. Clean the workpiece, removing any trace of rust, grease, dust etc.

3. Check the plate for damage or defects.

4. Apply a thin coating of marking ink to the surface and allow it to dry.

5. Using an engineer's square, check that the workpiece is true and square.

6. With the square, an engineer's rule and a scriber, mark a line down the centre of the workpiece (line datum).

7. Using dividers, set the centre distance to 100 mm using the graduations on the engineer's rule.

8. Measure a distance (greater than 25 mm) along the line datum from one end. Using a dot punch, mark the centre of the first arc (point datum).

9. Locate the divider in the dot mark and strike an arc of 100 mm.

10. Use the dot punch to mark where the arc crosses the centreline, then repeat the process to produce two arcs marking the other two centres (point datums).

11. Set the dividers to 10 mm using the engineer's rule and scribe the 20 mm diameter circles.

12. Set the dividers to 25 mm and, using the dot marks in the centre of the 10 mm circles, scribe the 25 mm radius arcs.

13. Using the centre dot mark, scribe the 50 mm diameter circle.

14. Set the dividers to 50 mm using the engineer's rule and, using the centre dot mark, scribe the arcs representing the widest points of the component.

15. Using the engineer's rule, scribe four tangential lines connecting the 50 mm radii and the 25 mm radii.

16. Use a centre punch and the dot marks representing the centres, to enlarge the dot marks in preparation for drilling.

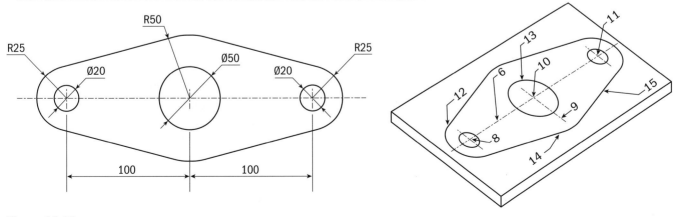

Figure 18.12

Marking out on a circular object

You need to take a different approach when you are handling and marking out round bars or cylindrical objects.

Activity: Marking out round bar

Select a piece of round bar and try the following marking out activities.

1. Use a vernier calliper and a rule to check that the round bar is large enough to allow the component to be produced.

2. Clean the workpiece, removing any trace of rust, grease, dust etc.

3. Apply a thin coating of marking ink to the end surface and allow it to dry.

4. Support the round bar in vee blocks.

5. Using a scribing block, or a box square and scriber, scribe a horizontal line along the length of the bar (line datum).

6. Using a combination square, with a centre head, scribe a line across the end diameter of the bar (line datum).

7. Rotate the combination square and centre head and scribe another line across the end diameter (line datum).

8. Clamp the round bar in place and use a dot punch to mark the intersection of the two lines (point datum).

Marking out on an irregular shape

A 3D object such as a casting or forging is more complex to mark out. As with the previous methods, you should use a rule to check that the workpiece is large enough to allow the component to be produced. Then clean the workpiece removing any trace of rust, grease, dust etc. For a casting, instead of using marking ink it is usual to apply a thin coating of whitewash or paint to the surface and allow it to dry.

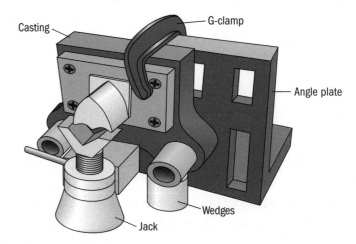

Figure 18.13: Using jacks and packing pieces to support an awkwardly shaped casting

You should use jacks and packing pieces to support the complex workpiece on the surface table. Once the casting is supported, you can clamp it to an angle plate to ensure that it is securely fixed. You can now use a scribing block or vernier height gauge to mark datum lines on the casting.

Because the surface table is acting as your datum, you can use a scribing block, engineer's square, combination square or rule to mark out lines. And because the workpiece is probably an awkward shape you will probably need to use a wider variety of techniques, involving slip gauges, DTIs etc.

18.2.3 Safe working practice

Marking out activities take place in an engineering workshop. There are risks associated with this type of environment and you should take care at all times. Health and safety is of prime importance and your tutor/assessor will give you guidance on the expected behaviour and normal working practices in the workshop.

Personal protection and hygiene procedures

While you are in the workshop you should wear protective clothing and protect yourself from danger or risks to health. Although you may be involved in marking out activities it is likely that you will be working close to machines, or passing by them and by other people who are working at the machines.

You should wear overalls or a boiler suit. Ensure that this fits correctly, that all buttons and zips are securely fastened and that your hair, if long, is tied back.

Things to avoid:

- Tools and equipment in pockets – these can cause cuts or harm to others and can also produce holes in pockets.

- Too small – overalls that are too small can restrict movement, or leave parts of the body unprotected.

- Too long – overalls that are too big can cause you to trip and fall, or get caught in machinery.

- Missing buttons/zips – anything that is not secure can come loose, or get caught in equipment or machines.

- Exposed jewellery or clothing – this should be removed or tucked out of the way.

Footwear is important. When you walk through a workshop you might walk through chemical spills, step on sharp pieces of metal or drop objects on your foot. If you wear ordinary shoes there is a real danger that severe damage can be caused to your feet. For these reasons you should always wear safety shoes or safety boots. These have non-slip soles and are made from a material that can resist oils and chemicals. The most likely accident to occur when marking out is that a heavy or sharp object may fall from the marking out table, so safety boots are lined with steel to ensure that if a heavy object does drop onto your foot you won't suffer from crushed or broken toes.

Whenever you are in a workshop you should always consider eye protection. Goggles, visors and other items of eye and face protection are used for specific workshop activities. For general use you should be provided with safety glasses, which you should wear at all times. Safety glasses have plastic lenses that are shatter resistant; they also have side protection to help prevent dust and dirt getting into the eyes.

You should wash your hands rigorously, using soap and water, before you enter a workshop and after you leave. Also, before you enter the workshop or do any marking out make sure you rub barrier cream into your hands. This provides a physical barrier between your skin and contaminants such as grease and oils. During marking out you will be handling workpieces, which will involve coming into contact with substances that can cause skin irritation or dermatitis. Gloves are provided for specific applications; if you are lifting or handling difficult or heavy objects you should wear the gloves that your tutor or assessor will provide.

Behaviour in the workshop

In a workshop environment there is a real danger that any foolish behaviour can cause serious injuries. Many workshop activities require focused concentration; shouting, pushing or playing games can distract an operator and cause danger to them and to others.

Avoid the following activities:

- shouting
- pushing
- horseplay
- practical jokes
- throwing objects
- playing games.

Many employees in industry have been injured or even killed by getting trapped in machines. Other risks include:

- electric shock
- knocking over heavy equipment or shelving
- being hit by forklift trucks or other vehicles
- contact with compressed air or hydraulic fluid
- falling from height.

Remember

Remember you should never be in a workshop without a tutor, assessor or appropriate supervisor.

Maintaining a safe and tidy work area

Maintaining a tidy workshop is an important aid to health and safety and can prevent serious accidents. Untidy workplaces are more likely to be unsafe workplaces. In fact, slips and trips are the most common cause of accidents in the workplace.

Tidy workshop checklist

- Doorway and entrances – Is there any obstruction, or water on the floor, making it slippery

- Trip hazards – Are there any cables, mats with curled-up edges or objects that could cause trip hazards?

- Contamination – Are floors dusty, or contaminated with oil or other liquids?

- Water – Is there water on the floor from a leak or broken window?

- Walkways – Are all walkways clear of boxes, equipment and clutter?

- Tools – Have all tools been returned to the correct storage area?

- Components – Are all workpieces correctly stored before and after being worked on?

- Hazardous substances – Are all hazardous substances safely secured and locked away when not in use?

18.2.4 Housekeeping

In a workshop environment, housekeeping is the process by which we ensure that we clean and store away all tools, equipment and other resources before we leave.

A place for everything and everything in its place

If you have to look through a toolbox or search the workshop for a particular tool, you can waste a considerable amount of time. If a shadow board is used, you can easily see where a tool is and where it should be kept. Of course, if people don't return the tool to the shadow board when they have finished with it, the problem remains.

Many workplace organisations use the **5S** principle to organise and maintain a safe, clean and tidy workplace.

Key terms

5S – a method for organising a workplace. It is based on a Japanese principle. The five S-words can be translated as follows:

Seiri **(sort)** – Sort out all tools and equipment and discard those that you don't need.

Seiton **(set in order)** – Arrange tools and equipment close at hand, using shadow boards, for example.

Seiso **(sweep)** – Keep the workplace and tools and equipment clean and neat. At the end of the session make sure everything is returned to where it belongs.

Seiketsu **(standardise)** – Ensure activities are carried out in a standard, repeatable way. This is important in manufacturing operations.

Shitsuke **(sustain)** – Perhaps the most important element is to keep things neat and tidy once the work area has been sorted out and not let bad habits build up.

Cleaning of equipment

As well as ensuring that tools and equipment are put away properly in the correct storage container, it is important to make sure they are properly cleaned. Many of the high-precision tools and instruments used in marking out have to be carefully looked after to maintain their effectiveness. For example, slip gauges have a protective coating of grease or oil to prevent them from corrosion or contamination. This surface protection is wiped off in use and the gauges should be re-oiled or greased and returned to their storage box after use.

Disposal of waste

Workshop activities can generate waste materials. Much of this waste can be recycled, such as used coolant and the swarf (metal cuttings) generated during secondary machining operations. During marking out it is important to dispose of waste material carefully. For example, if paper towels are used to clean workpieces in preparation for marking out, they could be a fire hazard: they should be disposed of and not left in the workshop.

Storage of measuring and marking out equipment

In addition to cleaning measurement equipment and lightly oiling or greasing it, you should always return it to its storage box. Slip gauges should be in individual compartments and should never be in contact with each other in storage. Similarly, vee blocks are supplied in matched pairs and these should always be stored together in their box.

The storage boxes supplied with marking out equipment are designed to support the equipment to prevent distortion. Storing marking out tools with other tools can cause burrs or chipped edges, particularly to the sharp tips of scribers and dividers. All boxes should be stored in a dry environment, such as a cupboard or cabinet.

These slip gauges have been stored correctly in individual compartments

BTEC ## Assessment activity 18.3    P3 P4 P5 M2 D1

In Assessment Activity 18.2 you selected measuring and marking out equipment in order to complete three marking out activities.

In this assessment activity you are going to complete three different marking out activities supplied by your tutor/assessor. These will consist of a drawing or written instruction indicating the shape, size and key features of the desired workpiece. You should also have three different pieces of material or blanks.

1. You will be required to carry out a marking out activity using:

   - a square/rectangular application

   - a circular/cylindrical application

   - an irregular workpiece such as a casting or forging.

   You may well be able to use many of the tools and techniques you described in Assessment Activity 18.1. However, you will need to draw up separate work plans for each activity.

2. Draw up a flow chart showing the marking out sequence and explaining the marking out method you would use for each activity.

   You will be observed carrying out this activity and you will need to demonstrate and explain that you can work safely and use the tools and equipment properly.

3. In addition, you should ensure that you check that the marked out component meets the requirements of the written instructions or drawings.

4. Finally you should explain to your tutor/assessor:

   - why you selected and used your chosen work-holding devices

   - why you selected and used your chosen measurement tools and techniques during the marking out exercise.

   - what made you decide which datums to use.

PLTS

By drawing up your own work plan and organising your own tools and equipment you can demonstrate your **self-manager** skills in organising time and resources and prioritising actions.

There are risks in all workshop activities. By showing that you can work safely and manage any potential risks or accidents, you can demonstrate your **self-manager** skills in anticipating, taking and managing risks.

By ensuring that you tidy the workspace, clean and lubricate all tools and equipment as necessary and manage risks, you can demonstrate your **effective participator** skills in identifying improvements that would benefit others as well as yourself. By improving the workplace you will benefit all your colleagues.

 ## Functional skills

When you use engineering drawings to determine sizes and features to be marked out on the workpiece you will be developing your **English** skills in comparing, selecting, reading and understanding texts and using them to gather information, ideas, arguments and opinions.

When you are describing the work-holding and marking out equipment and justifying the choice of datum, work-holding equipment and measurement techniques, you will be demonstrating your **English** skills in writing documents, including extended writing pieces, communicating information, ideas and opinions, effectively and persuasively.

BTEC Assessment activity 18.3   P3 P4 P5 M2 D1

Grading tips

P3 It is important to be able to plan the marking out procedure, because you will need to consider the order and sequence in which the workpiece will be moved, clamped and adjusted. To achieve this grade you will need to demonstrate that you can prepare a work plan for three different marking out activities.

P4 It requires patience and a plan to mark out different types of workpiece. Make sure you follow the step-by-step procedure carefully for each workpiece. To achieve this grade you will need to demonstrate competence in carrying out different marking out activities.

P5 You should ensure that when carrying out this activity you clean all tools and equipment and return them to their storage boxes when not in use. Keep your workplace clean and tidy and work safely. To achieve this grade you will need to demonstrate safe working practices and good housekeeping procedures.

M2 To ensure that the marking out activity you have completed is fit for purpose, you should make careful measurement checks with reference to the drawing or written instructions. To achieve this grade you will need to carry out checks to ensure that the marked out components meet the requirements of the drawing or job description.

D1 Justifying the use of tools and techniques requires you to explain why you used certain techniques rather than others. To achieve this grade you will need to justify the choices of datum, work-holding equipment and measurement techniques used to mark out the three different applications.

JANSHER KHAN

Skilled machinist

I work in a small workshop where we manufacture components for Formula 1 racing cars.

We are a small team and each of us uses a variety of the machines in the workshop. We mark out our own workpieces and then carry out the secondary machining operations. The work is very interesting, as we have to manufacture very small numbers of components, or even individual items. The race teams are always testing different components, designs and materials, so it can be quite challenging.

Typical day

A typical day starts with the team getting prepared: putting our overalls on and rubbing barrier cream into our hands. We then discuss with the racing engineer which jobs are a priority and the engineer hands out work plans and drawings. Each day is different, depending on the parts of the car we are working on. It also depends on whether the race team are testing or need new parts during the racing season. I usually work on the grinding machine, but there can be days when I use the lathe or the milling machine. We spend a lot of time measuring and checking, because we need to work to very close tolerances and this needs a high degree of accuracy. We are responsible for marking up our own jobs, so it is important that I make a good job of this, as I shall have to do the machining afterwards. The teamwork is great and we all try to help each other to make the best-quality components possible.

The best thing about the job

I really enjoy working with unusual materials, which can be very challenging. The high precision and accuracy of the work are things that are very important to all of us. It's great to watch the testing and we all celebrate when the team wins a race. The parts we manufacture have helped the car to win.

Think about it!

1. Think of the activities that you have completed in this unit. What sort of components would you be manufacturing for a racing car?

2. Racing cars have to be constructed from lightweight materials. What sort of materials might they be and how would you mark them out?

3. If the car has one part that needs marking out in the same way many times, what could help make this exercise quicker and easier?

Just checking

1. Describe three different tools used for scribing lines on a workpiece.

2. Describe three different ways that a workpiece can be secured before marking out.

3. Describe how three different measuring tools are used.

4. Explain the reasons for drawing up a work plan before marking out components.

5. Explain why good housekeeping in the workplace is a health and safety issue.

Assignment tips

When you are selecting the correct marking out technique for a given workpiece, there are several questions you need to ask yourself.

- What kinds of datum are already present on the workpiece, if any?

- Is the workpiece to be mounted horizontally, vertically, or at an angle?

- Is there an engineering drawing to work to?

- Which technique will give you the best accuracy?

- Can you carry out the operation safely yourself, or will you need assistance?

- Will the workpiece require several operations on different machines?

You might find it useful to complete a work plan for the marking out activity. This is a step-by-step guide to carrying out the operation. You can specify the marking out equipment and techniques you will need to complete the activity.

19 Electronic circuit construction

To work as an electronics technician you need a variety of skills. You need to understand electronic theory, so that you can select the correct components for a given function. You also need to be able to read and understand circuit diagrams and know how to construct electronic circuits using various construction techniques. It is equally important that you do this safely and are aware of the hazards that could occur in the workshop.

This unit will introduce you to all of the above. First you'll gain an understanding of the safe working practices that you need when you're working with electronic components and circuits. You'll learn about the hazards and risks that can occur when constructing electronic circuits in a workshop or laboratory. You'll then move on to develop an understanding of the function of various electronic components and how they are represented in circuit diagrams.

Once you've mastered electronic component theory and can read circuit diagrams, you'll progress to learning how circuit boards are manufactured and constructed. You'll be introduced to the various methods used and learn how to select the appropriate electronic components needed to build some complete circuits.

Learning outcomes

After completing this unit you should:

1. be able to use safe working practices in the electronics laboratory/workshop
2. know about electronic components and circuit diagrams
3. know about the manufacture of electronic circuit boards
4. be able to construct an electronic circuit.

Assessment and grading criteria

This table shows you what you must do in order to achieve a **pass**, **merit** or **distinction** and where you can find activities in this book to help you.

To achieve a **pass** grade the evidence must show that you are able to:	To achieve a **merit** grade the evidence must show that, in addition to the pass criteria, you are able to:	To achieve a **distinction** grade the evidence must show that, in addition to the pass and merit criteria, you are able to:
P1 describe the potential hazards related to constructing electronic circuits **Assessment activity 19.1 page 249**	**M1** explain the function and operation of four different electronic components **Assessment activity 19.2 page 260**	**D1** propose a method used to construct a given electronic circuit and justify your choice. **Assessment activity 19.3 page 264**
P2 use safe working practices in the electronics workshop/ laboratory **Assessment activity 19.1 page 249**	**M2** explain the advantages and disadvantages of the three types of electronic circuit board. **Assessment activity 19.2 page 260**	
P3 describe the purpose of six different types of electronic component **Assessment activity 19.2 page 260**		
P4 read a given circuit diagram to identify the electronic components in the circuit **Assessment activity 19.2 page 260**		
P5 describe the manufacture of the three types of electronic circuit boards **Assessment activity 19.2 page 260**		
P6 use two methods of construction for a given electronic circuit. **Assessment activity 19.3 page 264**		

How you will be assessed

This unit will be assessed by a series of assignments designed to allow you to show your understanding of the unit outcomes. There will be one assignment where you need to identify hazards related to working in an electronics workshop or laboratory. In a further assignment you will be required to identify and explain the purpose of given electronic components and describe three different methods of constructing an electronic circuit. A third assignment will allow you to demonstrate your practical skills by using two different methods for constructing a simple electronic circuit to a given specification.

Overall, your assessment will be in the form of:

- practical tasks
- written assignments
- oral questioning.

Peter, 16-year-old electronics apprentice

I've been interested in electronics ever since I built a small circuit in a lesson at school. I decided then that I wanted a job working with electronic circuits and this unit helped me understand the basic principles, plus how to be safe when working in an electrical or electronic workshop. The last thing I wanted was to get an electric shock, or leave a piece of equipment unsafe for someone else to use.

I enjoyed learning about different types of component, how they work and how to identify them. Most important of all, this unit taught me how to read a circuit diagram, something I definitely couldn't do before.

It really interested me to see how circuit boards are made, the types of boards you can use and the different circuit construction methods. I also learned how to solder correctly – very important if you want your circuit to work properly the first time you switch it on.

The best part of the unit for me was actually building a circuit and seeing it work for the first time. It meant that I had done all the work properly and it felt very satisfying. It gave me the confidence that I could do a job properly.

Over to you

- Are there any areas of this unit that you think you will find difficult?
- How do you think you can prepare yourself for working on this unit?

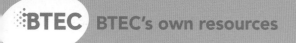

19.1 Be able to use safe working practices in the electronics laboratory/workshop

Start up

Can you spot potential hazards?

The electronics laboratory or workshop contains a lot of equipment and materials that could present a hazard if not used, handled or stored properly.

Working in groups of three or four, discuss the tools, equipment, substances and chemicals that you think you would find when working in the electronics laboratory or workshop. Then draw up a list of at least seven potential hazards. For example, you might run the risk of getting burned from an unprotected soldering iron.

Once you have done this, discuss your list of potential hazards with the other groups. See whether anyone has identified hazards different from those you have listed.

19.1.1 Hazards

When you're working in the electronics workshop or laboratory you'll be surrounded by many potential hazards. Some of them are obvious, but others are not so obvious. For example, you can easily see that the chemicals you use to etch a printed circuit board could cause a problem if they're spilled or aren't stored correctly, but electricity is invisible, so any associated hazard is more difficult to spot. You need to be fully aware of both the hazards and the potential health risks in your working environment.

All electrical equipment powered from the mains supply is potentially dangerous. Common potential hazards include the obvious one – electric shock – and the risk of smoke, fumes or fire caused by overheating equipment or cables.

The tools you need to use in an electronic workshop can also present hazards. For example, don't forget to watch out for that soldering iron: has it been put back into its holder properly?

Typical sources of hazards to be aware of are:

- soldering irons (e.g. burns, fumes, lead content of **solder**)
- sharp-edged hand tools (e.g. cuts, abrasions, swarf from cutting component leads, drills)
- **toxic substances**, chemical compounds and fumes (e.g. beryllium oxide, lead, solvents, etching fluid).

Key terms

Solder – a soft metal alloy of lead and tin. It has a low melting point and is used for joining electrical wires and components together. The solder contains flux, which helps to clean the surfaces to be joined and protects the joint once the solder has set.

Toxic substances – chemicals or materials that are poisonous and need to be handled correctly. Some may give off fumes, so should be used in a well-ventilated area; others could be corrosive to the skin.

Safe soldering

Soldering needn't be a hazardous operation if you take some simple safety precautions. The most obvious one is: don't touch the tip of the soldering iron with your fingers. It will be extremely hot (sometimes as high as 350°C), so accidental contact with it is definitely not desirable! Also, remember that the tip will burn, melt or even set fire to materials such as paper, plastic or cloths left lying around if it touches them. Use a stand for the iron in which the hot parts are properly protected and always put the iron back in the stand after use.

The lead content of solder used for electronic work is gradually being reduced, thus minimising this particular hazard. However, the flux in the solder gives off fumes, so make sure the work area where you are soldering is well ventilated. It is good practice to fit fume extraction equipment at the soldering station and to protect yourself from solder or fume damage by wearing eye protection, such as safety glasses.

Working with hand tools and chemicals

It's obvious that you need to take care when you're using hand tools, but the way you actually pick up and handle a tool is equally important. For example, a drill bit and the tool used to cut the copper track on stripboard, both have sharp edges. So these tools could cut you if you grab them, rather than pick them up carefully. You could also cut your unprotected hand while you're sweeping aside the swarf left over from drilling a piece of metal.

It's essential to handle chemicals safely: a stray splash whilst pouring a liquid; the slip of a container because your hands were wet; the strong fumes from certain chemicals – all these could have serious consequences. Handle chemicals as you would sharp objects. Wear gloves whenever possible; keep the work area and surface clean and tidy; and wear additional protective items as appropriate, such as glasses when there is a risk of damage to your eyes. Use a respirator when there are toxic substances involved, even if the area is well ventilated.

Figure 19.1: The skull and crossbones symbol is used to identify toxic substances

Remember

When solder touches the hot tip of the soldering iron, fumes are given off as the solder and flux melt. You should therefore always carry out soldering in a well-ventilated area.

Did you know?

Although very thin, the solder you use in the electronics workshop actually has flux running through it, rather like the letters running through a stick of seaside rock.

 Activity: Toxic substances

Work in pairs.

Use the Internet to research different toxic substances or materials that you might find in an electronics laboratory or workshop. Make sure you familiarise yourself with their properties, uses and effects. Then list your results in a table like the one below.

Name	Type of substance	Used for	Precautions
Ferric chloride	Liquid	Etching circuit boards	Corrosive; avoid skin and eye contact

Now compare your table with that of the other groups. See if you can work out which are the most common substances and write down which you think is the most dangerous and hence requires the most care when handling.

19.1.2 Safe working practices

Any workshop is a dangerous place if you don't take proper care while you're working in it. So, as well as being able to identify hazards or potential hazards in the electronics workshop or laboratory, you also need to be familiar with any rules, regulations and procedures associated with your working environment.

There are many do's and don'ts that you should be aware of. Procedures for checking that equipment is safe before you use it, regulations for how it should be used, the need to wear personal protective equipment – all these come under the heading of *safe working practices*.

You need to be familiar with:

- how to use hand tools safely when constructing electronic circuits (e.g. drills, pliers, knives, scalpels, screwdrivers, wire cutters and strippers, soldering irons): this was covered in detail in Unit 1
- the correct handling and storage of components and test equipment
- when to use personal protective equipment (PPE) such as safety glasses
- the cable colour-coding convention for mains equipment
- the correct procedure for checking earth connections

- how to correctly select and fit a **fuse** for a device of known power
- the safe replacement of a mains plug on a three-core cable
- **polarity** issues and the dangers or consequences of incorrect polarity (correctly connecting equipment or a circuit to a power supply, the correct polarity for cells and batteries, electrolytic capacitors and semiconductor devices).

Cable colour coding

Mains electrical equipment is connected to its power source by means of a cable containing a number of wires. These wires are colour-coded to help you identify which wire is which: see Table 19.1.

Table 19.1: The colours of inner wires within a cable

Colour	Wire
Blue	Neutral
Brown	Live
Green and yellow stripes	Earth

Wiring the mains plug

To connect the cable to the mains supply, it needs a correctly wired plug on the end. When you replace a mains plug on a cable, always ensure that you have wired the blue wire to the neutral terminal, the brown wire to the live terminal and the green and yellow striped wire to the earth terminal. Also, make sure that the cable is securely held in place by the cable grip; you don't want it to work loose during use and become a hazard.

Fuses

The mains plug contains a fuse. Its job is to 'blow' if its current rating is exceeded and hence break the supply of mains electricity to the equipment. It therefore helps to protect the equipment from possible damage or even fire if the equipment or the cable overheats because excessive current is being drawn.

Always fit or replace the fuse with the correct type. Don't simply assume that if, for example, a 3 Amp fuse has been fitted in a plug then it should be replaced by another 3 Amp fuse if it blows. Always check the power rating of the equipment it is supplying when you fit or change a fuse. Table 19.2 shows typical power ratings and which fuse you should use.

Table 19.2: The correct fuses for typical power ratings

Power rating	Fuse to use
Up to 700 W	3 Amp
Between 700 W and 1200 W	5 Amp
Over 1200 W	13 Amp

Key terms

Fuse – a thin wire conductor designed to fail and thus protect a circuit when the current passing through it exceeds a prescribed value. The fuse is said to 'blow', because the higher current causes the fuse wire to melt, thus breaking the supply to the circuit.

Polarity – the direction of electron flow from a power source such as a power supply, a cell or a battery. Electrons flow from the negative terminal through the circuit to the positive terminal. It is therefore important to connect a circuit using the correct polarity.

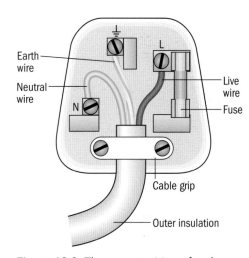

Figure 19.2: The correct wiring of a plug

Standard mains plug fuses

Polarity

It is essential that you observe the correct polarity when you connect electrical equipment or an electrical component to its power source. This means making sure that the neutral, live and earth wires are connected to the correct terminals, or that batteries are inserted correctly so that the positive terminal is connected to the positive supply line and the negative terminal is connected to the negative line.

Incorrect polarity could damage the equipment and make it unusable. If the power source is a power supply then incorrect polarity could cause it to fail, or overheat or even catch fire. Of course, any protective fuse should prevent the worst happening, but only if it is of the correct rating.

Always take special care with mains-operated equipment. An error in the wiring to the power source could make the equipment 'live'. As a result, instead of being at earth potential, the equipment is at live potential instead – that is, at 230 V a.c. Touching the equipment would result in an electric shock, so always check the wiring carefully.

It doesn't take much current to produce a serious electric shock on the human body. Other factors that affect the severity of the shock are whether the current is alternating or direct and the path it takes travelling through the body.

Table 19.3 shows the typical effects of different levels of current on the human body, assuming that the current source is a 50 Hz, 230 V a.c. mains supply.

Table 19.3: The effects of an electric current on the body

Level of current	Effect
Less than 1 mA	Hardly noticeable
1 mA to 2 mA	Threshold of perception – a slight tingle may be felt
2 mA to 4 mA	Mild shock – effects of current flow definitely felt
4 mA to 10 mA	Serious and painful shock felt
10 mA to 20 mA	Nerve paralysis occurs – unable to let go
20 mA to 50 mA	Breathing may stop – loss of consciousness
Greater than 50 mA	Burns and probable heart failure

When you are constructing electronic circuits, it is important to observe the correct polarity when you connect the components. Some electronic components, such as simple resistors and capacitors, are polarity independent, but electrolytic capacitors and semiconductor devices, such as the transistor, are polarity dependent. They could fail, become extremely hot and result in a burn hazard, or even explode if the correct polarity hasn't been observed. The simple rule is: always check and then re-check your work to avoid unnecessary problems and dangers.

You should also familiarise yourself with the correct procedure for dealing with any situations that might arise. To learn the first aid procedure for treating someone who has suffered an electric shock and how to treat electrical and acid burns go to the hotlinks section on page ii.

If you look around the workplace you should see, prominently displayed, posters or notices that show both how to avoid and, if necessary, how to deal with such incidents. Learn these procedures: you may well save your own or your workmate's life.

Electric shock procedure

In the event of an electric shock to someone you should do the following:

- If it is safe to do so, immediately switch off the electricity supply.

- If it is not possible to switch off the electricity supply, drag the affected person clear using some form of insulating material that may be to hand, such as a *dry* piece of cloth or any plastic material. On no account should you touch the person with your bare hands, as you risk being electrocuted yourself.

- If the person has stopped breathing and you have received the appropriate training, administer artificial respiration immediately.

- If you know how to feel for a pulse, but cannot detect one, then it means that heart massage will be required. If you do not know how to do this then you must send for a trained first aid person *immediately*.

- If you need help, send someone else nearby to get it. Don't wait for help to come to you or go looking for it yourself.

Fire procedure

You need to know what to do if you discover a fire:

- Raise the alarm immediately; then call the fire service, or get someone else to do it for you.

- If you have to leave the room to raise the alarm, first close all doors and windows to prevent the fire from spreading.

- Once you've raised the alarm, start evacuating the premises. There will be a procedure for doing this, so make sure you are familiar with it and take note of where the fire exits and assembly points are when fire drills are held. Someone will be in charge of checking the various parts of the building and taking a head count at the assembly point to make sure that everyone has been evacuated.

- Make sure all fire doors are closed as you leave, to prevent the spread of smoke. Smoke is the biggest cause of panic in a fire, which in turn leads to accidents, especially on staircases. Remember to not use lifts!

Remember

It is not volts that kills but the level of current flowing through the body. There is an old saying that will help you remember this – it's the volts that jolts but the mils that kills (meaning milliamps).

Did you know?

Any voltage source in excess of 50 V should be considered dangerous. However, in some circumstances smaller voltages could also be dangerous. Get in to the habit of avoiding contact with electronic circuits and treat all of them with great care.

- If the fire is small, you can attempt to contain it yourself using the appropriate fire extinguisher, as long as you don't put yourself in danger by doing so. Always make sure you know how to get out quickly, just in case. Let the fire brigade take over when they arrive.

- Don't attempt to re-enter the building until you are advised that it is safe to do so.

Activity: Poster on dealing with electric shock

Produce a large A3 poster, suitable for displaying in your workshop, giving information on how to deal with electric shock. The poster should be eye catching and easy to read and understand. Make sure you include details of how and where to switch off the electricity supply and relevant emergency first aid contact information.

The following all present an electric shock hazard:

- a broken plug
- a damaged cable
- exposed connections

A portable appliance testing (PAT) instrument is used for checking electrical safety.

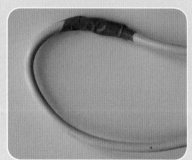

Some potential electric shock hazards: a broken plug; a damaged cable; exposed connections.

Key terms

Hazard – anything that may cause harm, such as electricity, chemicals, an unprotected soldering iron, or sharp tools.

Risk – the chance (high, medium or low) that somebody could be harmed by these and other hazards, together with an indication of how serious the harm could be.

Risk assessment: procedures for establishing risk

The Management of Health and Safety at Work Regulations 1999 require a risk assessment to be carried out in the workplace in order to identify any potential **hazard** or hazardous activity. Each identified **risk** is then assessed and graded. For example, you could grade a risk as follows:

- High – existing control measures are inadequate; immediate action is needed to reduce the risk.

- Medium – existing control measures are insufficient; existing measures are required.

- Low – no action is needed, as existing risk control measures are sufficient.

Assessment activity 19.1

Use the Internet to download and print out the leaflet *Five steps to risk assessment*. To access this website, go to the hotlinks section on page ii Familiarise yourself with this risk assessment leaflet so that you can carry out a meaningful and detailed risk assessment of the electrical/electronic workshop in your place of study. Produce a table listing all the risks you identify, the grade you have given to each risk (high, medium or low) and details of why you have so graded the risk. Also, write notes against each risk detailing what measures, if any, need to be taken to control that risk. **P1**

Using your risk assessment table, write a short report identifying which particular workshop activities call for the use of PPE (personal protective equipment) and what specific equipment this would be. **P2**

Grading tips

P1 **P2** As well as your Five steps to risk assessment leaflet, research risk assessment on the Internet and use the following questions to help you:

- Are all the electrical outlets in a safe condition?

- Does each item of equipment have a mains lead that is in a safe condition?

- Is each power lead fitted with a mains plug that is in a safe condition?

- Are all hand tools in a safe condition, especially those with sharp edges?

- Are tools correctly stored and not simply left lying around?

- Are work surfaces clear of wire cuttings or swarf from drilling?

- Does each soldering iron or soldering station offer adequate protection from burns or fumes?

- Are toxic substances correctly labelled and properly stored?

- Are safety notices prominently displayed? Do they list first aid procedures or an emergency contact?

PLTS

Producing a risk assessment report and identifying safe working practices will help you develop your skills as a **self-manager**.

Functional skills

Using the Internet for research and a word processing program to produce the report will help develop your **ICT skills**.

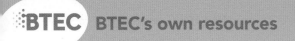
19.2 Know about electronic components and circuit diagrams

Start up

Types of electronic component

There are many types of electronic component, from the simple resistor to the self-contained integrated circuit. For each type of component there are usually different versions, depending on where and how it will be used.

Using the Internet, research the different types of the following electronic components:

- resistors
- capacitors
- semiconductors
- indicators.

Draw up a table like the one below listing each of the electronic components, the different types available, a brief description of each and where it would be used or what it would be used for. An example has already been entered for you.

Once you have done this, discuss your list of potential hazards with the other groups. See whether anyone has identified hazards different from those you have listed.

Component	Type	Description	Use
Resistors	Wire wound	Made of wire wound on a ceramic former	High-power circuit, or where an accurate resistance value is needed

19.2.1 Component types

Individual electronic components are connected together to make an electronic circuit. Each component has a specific function, so before you can design a circuit you first need to learn about the different types of component and what job they do.

For each component there is normally a variety of different types, each designed for and suited to a specific application. Table 19.4 shows some of the more common types.

Table 19.4: Some common electronic components

Component	Description	Types
Cell Battery	Primary cell – approximately 1.5 V, cannot be recharged. Secondary cell – approx 1.5 V, rechargeable. Power source supplying d.c. (**direct current**) energy. Made up of a number of cells to increase the power.	Alkaline, lead-acid, NiCd, NiMH.
Transformer	Coils of wire wound around a metal-based former. Input winding is the primary, output winding is the secondary. Can only work in an a.c. (**alternating current**) circuit; cannot work with d.c.	Step-up, step-down.
Inductor/choke	A conducting wire, usually in the form of a coil. Stores the energy in a magnetic field created by the electric current passing through it. Commonly used in fluorescent lamp and radio circuits. The unit of measure of inductance is the henry (H).	Air core, iron core, ferrite core, toroidal core, fixed/variable.
Indicator	Visual – to provide a visual indication of, for example, a power-on state. Audio – uses sound as an indicator. Sometimes used to provide an audible warning signal, such as a 'beep'.	Lamp, neon. Loudspeaker, piezo buzzer.
Resistor	Used to control the current flowing through a circuit. Available in a wide range of values and tolerances. The unit of measure of resistance is the ohm (Ω).	Carbon, metal film, wire wound, light dependent, fixed/variable.
Capacitor	Can be charged, so used to store electrical energy. Amount of charge depends on the capacitor value. Rate of discharge can be controlled. Used to smooth ripples in power supplies, for coupling and decoupling stages in a circuit, or tuning in a radio circuit. The unit of measure of capacitance is the farad (F)	Polyester, paper, Mylar, ceramic, electrolytic, tantalum, air-spaced, fixed/variable.
Diode	Allows current to flow in one direction only. Used to rectify a.c. and change it into d.c., for voltage stabilising, for signal detection in radio circuits and as an indicator.	Germanium, silicon, **zener**, light emitting (LED).
Transistor	A device for controlling and amplifying current. Used in amplifiers, switching and oscillator circuits.	PNP, NPN, field effect (FET).
Integrated circuit	Complete functional circuit on a single ship of silicon.	Op amps, logic gates, timers.

Key terms

Alternating current – often referred to as a.c.; electrical power that is not at a fixed level but alternates between a positive peak and a negative peak. The common waveform for alternating current is a sine wave. The mains supply is a.c.

Direct current – often referred to as d.c.; electrical power that is at a fixed level. Cells and batteries deliver direct current. Mains-powered battery chargers and adapters for small devices such as mobile phones and MP3 players contain a circuit to convert a.c. to d.c.

Zener – a diode that makes specific use of what is called the reverse bias effect. It can be designed to begin conducting at a specific reverse bias voltage and so is used for stabilising or regulating a voltage source.

Remember

Most electronic components are designed to function at certain power levels or voltage. When designing a circuit always make sure you choose a component that is up to the job and will not fail at some point because its power rating is too low.

Did you know?

Alternating current is both easier and cheaper to produce than direct current. It is also easier to distribute, as the voltage can be conveniently changed using transformers. This is why the mains supply is alternating current.

A variety of electrical components mounted on a PCB

Component colour codes

Some components have their value printed on them, but others, such as resistors, are often identified using colours. Figure 19.3 shows the four-band and five-band resistor colour codes and how to use them.

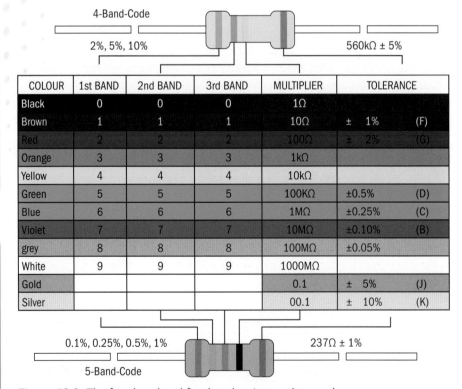

COLOUR	1st BAND	2nd BAND	3rd BAND	MULTIPLIER	TOLERANCE	
Black	0	0	0	1Ω		
Brown	1	1	1	10Ω	± 1%	(F)
Red	2	2	2	100Ω	± 2%	(G)
Orange	3	3	3	1kΩ		
Yellow	4	4	4	10kΩ		
Green	5	5	5	100KΩ	±0.5%	(D)
Blue	6	6	6	1MΩ	±0.25%	(C)
Violet	7	7	7	10MΩ	±0.10%	(B)
grey	8	8	8	100MΩ	±0.05%	
White	9	9	9	1000MΩ		
Gold				0.1	± 5%	(J)
Silver				00.1	± 10%	(K)

Figure 19.3: The four-band and five-band resistor colour codes

From Figure 19.3 you can see that, for a four band resistor, green–blue–yellow–gold identifies a 560 kΩ resistor with a tolerance of ± 5%.

For a five-band resistor, red–orange–violet–black–brown identifies a 237 Ω resistor with a tolerance of ± 1%.

Activity: Resistor colour codes

Try this on your own.

Using the resistor colour code chart in Figure 19.3, write down the values of the following resistors:

- brown red brown gold
- orange orange orange silver
- red black green brown
- blue grey orange red
- yellow violet green green
- orange red yellow orange brown
- red yellow violet blue green
- green blue brown black blue.

Now work out the colour codes of the following resistor values:

- 1 kΩ ± 10%
- 4.7 kΩ ± 2%
- 330 Ω ± 5%
- 0.22 Ω ± 1%
- 249 kΩ ± 1%
- 3.62 MΩ ± 0.5%

19.2.2 Circuit diagrams

A circuit diagram shows how the various components of an electronic circuit are connected together. Standard symbols are used for the components and interconnections or linkages, so the diagram is easy to read and understand once you have mastered the symbols. The symbols, with a few small exceptions, are like a universal language, because an electronics engineer in one country should have little difficulty in understanding a circuit diagram produced by an engineer in another country.

Component and interconnection symbols

You need to become familiar with the standard electronic symbols before you can understand a circuit diagram. The most commonly used standard symbols are shown in the Appendix. As you can see there, the symbols bear little resemblance to the actual component; they are simply used to represent the component.

These particular standard symbols have not always been used. You may well come across a circuit diagram in which the occasional component is represented by a different symbol. For example, the resistor may be shown as a zigzag line. American circuit diagrams often use this zigzag symbol. This is because the zigzag was the original symbol for a resistor and some designers still use this symbol rather than the standard one. Fortunately, the standard symbols are becoming more common.

There is also a standard for how to draw connections and linkages on a circuit diagram. This is also shown in the Appendix, as well as the correct and incorrect ways to draw lines that simply cross on the circuit diagram without linking.

Study the diagrams of components in the Appendix until you are able to identify all of the standard symbols without having to look them up. Do the same with the diagrams of connections and linkages; it shouldn't take you too long.

Activity: Standard electronic symbols

Work in pairs.

Study and learn the standard electronic component symbols shown in the Appendix. Test each other's ability to recognise a component correctly from its symbol by each drawing five randomly chosen symbols on a sheet of paper, then swapping sheets and naming the symbols. Hand the sheets back to each other, mark the answers, then repeat the exercise a number of times.

Now try doing the above activity again. This time write down the names of five components, swap sheets and draw the symbols for the components named. Hand the sheets back to each other, mark the answers, then repeat the exercise a number of times.

Keep doing this until you are very familiar with both the electronic components and their symbols.

There is also a convention for laying out a circuit diagram. For example, the circuit input should generally be on the left of the diagram and the output should be on the right. Similarly, when drawing the supply to the circuit, the ground or 0 V line is normally shown at the bottom of the diagram and the positive supply voltage goes at the top.

Not all circuit diagrams can follow this convention, especially if a circuit has more than one supply voltage, or many inputs and outputs. Also, a circuit often has many components connected to the ground or 0 V line. To simplify the drawing of the circuit diagram and avoid having to show too many wires all connecting to the same point, the connections may be shown using the appropriate symbol found in the Appendix.

A very simple circuit diagram is shown in Figure 19.4. Can you work out what it does and redraw it so that current flows around the circuit? (Hint: you will have to close the switch.)

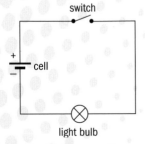

Figure 19.4: A simple three-component circuit

Component labelling

There is also a convention for labelling the individual components in a circuit diagram, usually starting from the left of the diagram and working across to the right. Table 19.5 lists a common convention for component labelling.

Table 19.5: Common component labelling.

Component	Label
Resistor (fixed)	R
Resistor (variable)	VR
Capacitor	C
Transformer	T
Inductor	L
Choke (radio frequency)	RFC
Diode	D
Zener diode	ZD
Transistor	TR

Did you know?

Some early electronic symbols used complex shapes, but many component symbols used today have been redesigned using a combination of straight lines and square or oblong shapes, making them easy to draw when using a computer and CAD software.

Activity: Label components on a circuit diagram

Try this on your own.

Draw the three-transistor circuit diagram shown in Figure 19.5, carefully and neatly, on a sheet of paper. Now label each component in the circuit, using the convention shown in Table 19.5 to help you. For example, the first resistor will be R1 and the first transistor will be TR1. Can you work out what is connected at the output?

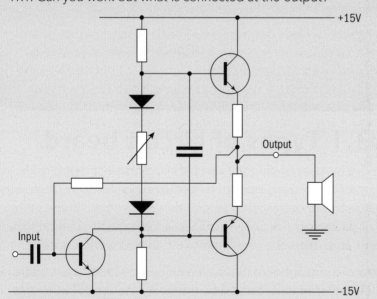

Figure 19.5: A three-transistor amplifier circuit

Block schematic

For circuit diagrams where various parts of the circuit have different functions, it is sometimes more convenient to also draw a block schematic or block diagram of the circuit. Each block in the diagram shows a different function within the circuit and the whole diagram shows how the blocks interconnect. Figure 19.6 shows the block schematic for a basic radio receiver circuit.

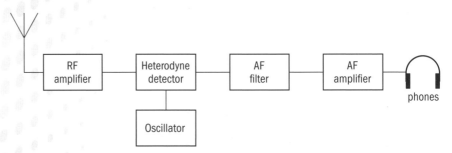

Figure 19.6: Block schematic of a basic radio receiver circuit

19.3 Know about the manufacture of electronic circuit boards

Start up

Types of circuit board

An electronic circuit can be built in a number of different ways. Use the Internet to research at least three different types of circuit board used in electronic circuit design and construction. For each, make a set of notes that you can use for reference while working through this learning objective.

Your notes should include a clear description of each type of circuit board, how you would use it to construct a circuit and when you would use this type of board.

Key term

Etched – The PCB starts life as a board with a coating of copper on one or both sides. The interconnecting wires are traced onto the copper using etch resist – something that will protect the copper and stop it being removed. The board is then placed in an etching solution that removes all the unprotected copper, leaving only the copper tracks. This process is called etching.

19.3.1 Types of circuit board

There are a number of different ways of building electronic circuits, but they all will involve using a board to fix the components to. The method you decide to use depends on several things. If you are at the early stages of designing a circuit you will use a fairly simple method that allows you to change the circuit easily if it doesn't work properly.

Your idea of a circuit board may well be a neat printed circuit board (PCB). Copper tracks are **etched** on the board in place of wires and the components are mounted on the board and soldered directly to the appropriate track. Very few, if any, wires are used and the whole assembly is very easy to manufacture on a large scale.

However, this is usually only after the circuit has been tested and passed for production. Before that the circuit will have been built in a very simple – and often untidy – manner. You need to use a method in which the components and wiring can be easily changed if the design is altered. Once the circuit has been tested, finalised and approved, you can design a PCB for it so it can be put into production.

Table 19.6 lists three common types of circuit board used in electronic circuit construction.

Key term

Prototype – the pre-production version of an electronic circuit or circuit board. The prototype is used for fully testing the final design or construction before the circuit is committed to mass production using a printed circuit board. You do not want to discover a problem with your circuit after the PCBs have been produced, so the prototype stage is an extremely important one if you want to avoid costly mistakes.

Table 19.6: Three common types of circuit board

Method	Use	Advantages	Disadvantages
Breadboard (or protoboard)	Used for the initial building and testing of new circuits. Not a permanent form of construction; components and wires are push-fitted or twisted together	• Changes can be made easily and quickly • Components can be reused	• Not suitable for permanent use • Not suitable for circuits with lots of components or more than about six active devices (e.g. transistors)
Strip and tag board	Used for building circuits at the **prototype** stage, before the circuit is put into production	• Cheaper and more permanent than breadboard as components are soldered in place • Board can be reused a number of times	• Copper tracks on the reverse of the board need to be cut with a knife • Care is needed to avoid short circuits while soldering
Printed circuit board (PCB)	Used for permanent circuit construction	• All the connecting wires are etched onto the board; no loose wires • Can be assembled by machine for mass production	• Difficult to modify if changes are made to the circuit • Difficult to change failed components; quicker to replace the complete board

A breadboard layout

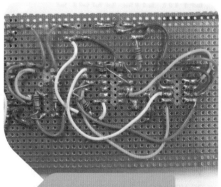

A typical stripboard construction (copper side)

A printed circuit board – note the tracks and component layout markings

PLTS

Being able to read a circuit diagram and identify the electronic components used in the circuit will help you develop your skills as an **independent enquirer**.

Functional skills

Using a word processing program to produce the report identifying the electronic components will help develop your **ICT skills**.

Grading tips

P3 These should be brief descriptions of why the component is being used.

M1 Understanding the function of different electronic components will help you become more efficient in circuit design.

P4 Ensure you include at least 6 components.

P5 The 3 methods are a breadboard, a stripboard and a printed circuit board. The 3 types of circuit board which you should consider are breadboard or protoboard; strip board or tag board; printed circuit board.

M2 Write a short report in which you explain the advantages and disadvantages of the three types of electronic circuit board construction that you described in P5. Try to illustrate your report with images.

BTEC Assessment activity 19.2   P3 M1 P4 P5 M2

The company you work for designs and constructs electronic circuits and has recently employed some new apprentices. You have been asked to produce information on electronic components, circuit diagrams, circuit design and construction for these new starters so that they can become familiar with electronic circuit construction as part of their training.

Your employer has also asked you to include details on building and testing circuits using standard methods. You must therefore produce an accurate and detailed report as follows.

1. Describe the purpose of the following electronic components:

   a resistor

   b capacitor

   c battery

   d transformer.

   e LED

   f LDR **P3**

2. Explain the function and operation of four different electronic components which your tutor will give you. **M1**

3. Identify and label the various electronic components you see in the circuit diagram shown in Figure 19.7 and then produce a simple components list. **P4**

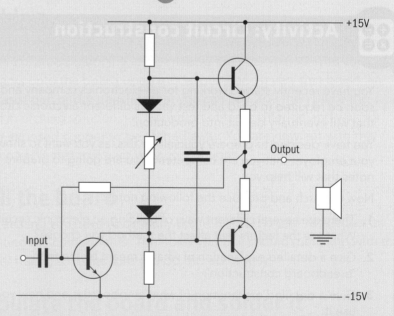

Figure 19.7

4. Describe the manufacturing processes used when constructing electronic circuits on three different types of circuit board. **P5**

5. Explain the advantages and disadvantages of working with a breadboard, strip board and printed circuit board. **M2**

19.4 Be able to construct an electronic circuit

Surface-mount technology

So far we have worked with types of component that have been around for years. However, there is another type of component that has become popular more recently, called a surface-mount device.

Work in groups of two

Use the Internet to research surface-mount technology and produce a short but accurate report comparing and contrasting the differences between surface-mount devices and standard electronic components. Complete the report by explaining the difference between a normal PCB and a PCB for surface-mount components. List two advantages and two disadvantages of using surface-mount components.

19.4.1 Circuit construction techniques

One of the secrets of successful circuit construction is good soldering. You don't just switch the soldering iron on, apply the tip to the joint you want to make and add some solder.

You should use an iron fitted with the correct tip for the job. If it is too small you won't be able to apply enough heat for a good joint. If it is too large you run the risk of soldering more than just the joint, or of applying too much heat and damaging the joint and the components. Ideally, you should use a temperature-controlled iron to provide the correct heat and further protect your work.

Before soldering, make sure that all the surfaces to be soldered are clean, so that the solder makes a good joint. Make sure you also keep the soldering iron bit clean, by using a wet sponge to wipe it on and **tinning** it regularly.

Problems you will encounter when soldering are:

- Too much solder – Solder should flow around the joint. If the joint has a large blob of solder on it, this indicates that the solder has not flowed correctly and is probably not making proper contact with the surfaces to be joined.

- Too little solder – Although the solder has flowed, not applying enough solder means that it may not have reached some surfaces: thus not everything is making contact.

Key term

Tinning – applying a coating of solder to keep the tip of the soldering iron clean and free from oxide and to aid heat transfer.

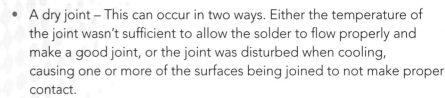

- A dry joint – This can occur in two ways. Either the temperature of the joint wasn't sufficient to allow the solder to flow properly and make a good joint, or the joint was disturbed when cooling, causing one or more of the surfaces being joined to not make proper contact.

A dry joint can also occur some time after the joint has been correctly soldered. Sometimes a joint is disturbed by the action of heating and cooling over a period of time, or by prolonged mechanical vibration. The soldered surfaces work loose and correct contact is lost. A dry joint often takes on a dull appearance, whereas a good sound joint should have a bright appearance.

Figure 19.8 shows what good and bad solder joints look like.

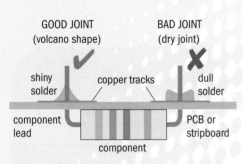

Figure 19.8: Good and bad solder joints

Using a heat sink

Certain electronic components, such as transistors, can be damaged by too much heat when soldering. Even if you are a soldering expert it is wise to use a heat sink clipped to the lead between the joint and the body of component. You can buy a special tool for this, but gripping the wire with fine-nosed pliers or a standard crocodile clip does the job just as well and is cheaper.

Wire wrapping

Wire wrapping is a solderless alternative to making a connection. It is normally used when terminating an external connection to a pin or post pushed through the circuit board. It can be made using a hand tool or small machine to wrap a piece of bare wire tightly around the post. Because it is not soldered, the wire is easy to remove should this become necessary.

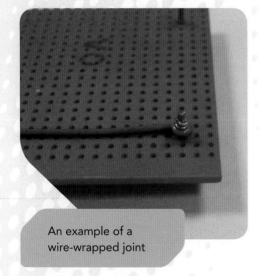

An example of a wire-wrapped joint

Labelling connections

It is always a good idea to label connections on your board to avoid making a mistake. For example, if you accidentally connect a power source to a sensitive input, it could have disastrous consequences for the component involved. Make sure you label all connections to and from the board and especially the power supply connections, inputs and outputs.

Electrical noise pickup

Another thing to consider when constructing a circuit is the layout of the connecting wires. In general, you should run wires in such a way that they minimise the risk of introducing electrical noise into the circuit. Keep connections wherever possible and don't route them close to sensitive components.

To help minimise electrical noise pickup, printed circuit boards often make use of a large area of copper etched onto the board to shield the circuit. This is called a ground plane, because all the earth and negative connections are made to it, so that it acts as a ground screen for the rest of the circuit.

Surface-mount technology

Surface-mount technology, or SMT, is the term for a method of construction in which the components have no long wire leads; instead they are mounted directly onto the surface of the printed circuit board using flat, tinned areas or contacts on the body of the device.

SMT components are therefore much smaller and lighter than their standard counterparts and are usually cheaper. The use of SMT can therefore result in lower cost, easier automation for mass board production and the ability to build quite a large circuit onto a relatively small PCB.

A selection of surface-mount technology components

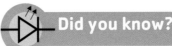
Did you know?

Some printed circuit boards not only have copper tracks etched on either side, but also have some running in layers through the middle of the board. Computer boards are often multi-layer like this. Boards with as many as seven layers are common and layers in excess of 20 are quite possible.

Key terms

Combinational logic circuit – logic circuit in which the output depends directly on the state of the inputs. Used for circuits where you know the output you require for known input conditions: using for example, NOT, AND, NAND, OR, NOR and EXOR gates.

Transistor – a device for controlling and amplifying current. There are two types (PNP and NPN). Both have three connections: base, emitter and collector. The field effect transistor (FET) also has three connections: gate, drain and source. All are used in amplifiers, audio, switching, optical and oscillator circuits.

Alarm circuit – circuit in which a single or combination of input conditions triggers the circuit into switching states. Used in intruder alarm systems, smoke alarms, car alarms and even low-battery indicators. These circuits use transistors, combinational logic or a mixture of both.

19.4.2 Types of circuit

There are many different types of electronic circuit. Typical examples include those using **transistors** (such as amplifiers, switching circuits and sensors), **combinational logic circuits**, **alarm circuits** and audio and optical circuits. Two simple circuits are shown in Figures 19.9 and 19.10.

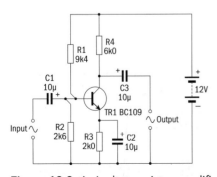

Figure 19.9: A single-transistor amplifier

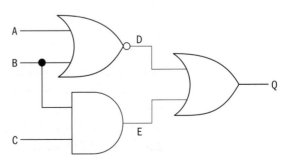

Inputs			Outputs		
A	B	C	D	E	Q
0	0	0	1	0	1
0	0	1	1	0	1
0	1	0	0	0	0
0	1	1	0	1	1
1	0	0	0	0	0
1	0	1	0	0	0
1	1	0	0	0	0
1	1	1	0	1	1

Figure 19.10: A combinational logic circuit and its truth table

Assessment activity 19.3

1. The circuit shown in Figure 19.11 is to be put into production. Propose which you believe will be the best method of construction for the circuit and adequately justify your choice. **D1**

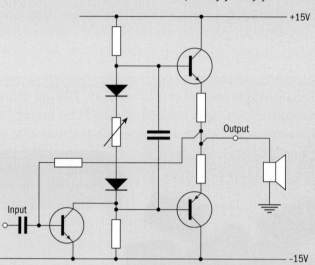

Figure 19.11:

2. Study the circuit shown in Figure 19.12 and design a suitable layout for it. Then construct the circuit accurately using:

a a breadboard

b a stripboard. **P6**

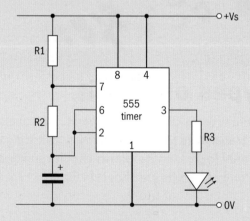

Figure 19.12: Timer circuit

PLTS

Constructing a given electronic circuit using two different methods will help you develop your skills as a **self-manager**.

Functional skills

Constructing electronic circuits will help develop your practical skills.

Grading tips

- Becoming familiar with the table of standard symbols will help you in identifying the various electronic components in a circuit diagram.

- Researching and becoming familiar with how to construct electronic circuits using different methods will help you become more efficient in circuit construction.

KATE STEVENS
Electronics engineer

I work for GM Electronics, constructing and testing electronic circuits. I am part of a team consisting of four engineers, each with their own specialist skills in a particular area of electronics and circuit design. There were very few female electronics engineers when I first started, but I've noticed that more are coming into the industry now.

I am responsible for:

- contributing to the design specification of a particular circuit
- designing the power supply and output sections of a circuit, as these are my own specialist areas
- producing accurate circuit diagrams
- building and testing prototypes of the circuit
- producing artwork for the production of printed circuit boards
- ensuring that the circuit design, building and testing go to schedule
- arranging for the circuit to go into production once it has passed testing
- ensuring all the relevant paperwork is completed.

My day starts with a meeting with the rest of the team to discuss progress on the design and prototyping of the latest circuits. We discuss any problems and how best to deal with them. It might be a part of the circuit that doesn't work when built on a stripboard and is failing its test routine, or a problem with designing the layout for a printed circuit board before it goes into production. I usually spend four hours a day on circuit design, sometimes using computer-based simulation software for the more complex circuits. I spend a further few hours building the circuit on a breadboard or stripboard, testing it to see if any modifications are needed, referring to other members of the team if necessary. The day is usually rounded off with a short progress report for my head of department.

I really enjoy being given an idea and coming up with an electronic circuit that does exactly what is required. After building the circuit it is a great feeling when I switch the power on and it works properly straight away. Mind you, sometimes it doesn't, so then I have to go back over the design and check the construction to see why. When I find the reason I feel pleased, because I know that the circuit now works properly.

Think about it!

1. What topics have you covered in this unit that provide you with the skills and background knowledge you need to become a good electronic circuit designer and constructor? Write a list and discuss it with your peers.

2. What further skills might you need to develop? For example, you might need additional training on producing artwork, or in the use of computerised design software in order to make your work more efficient. Write a list and discuss it in small groups

Just checking

1. Define 'combinational logic' and explain what an integrated circuit is.

2. What hazards can you identify with a soldering station?

3. What safety precautions should you observe when etching a printed circuit board?

4. Where would you use a Zener diode?

5. What types of transistor are there?

6. List the procedures for tending to someone who has just received an electric shock.

7. List five different types of capacitor and describe the differences between them.

8. What is a block schematic?

9. Why is it important to label input, output and power connections to a circuit board?

10. What is a multi-layer circuit board?

11. How many connections does a transistor have? Name them.

edexcel

Assignment tips

There are many ways to find information on the topics covered in this unit. You can use the library, but by far the most convenient way is to research thing on the Internet. Do remember, though, that not everything that finds its way onto the Internet is accurate. The following tips will help you get the best results:

- When you are using a search engine, think carefully about the key words you enter. For example, 'surface-mount technology' will produce more relevant results than simply entering 'surface-mount' – think about it.

- Don't automatically take as fact everything that you read, such as comments posted on discussion groups, online 'open' encyclopaedias, or explanations posted by individuals that have not been verified. There is a lot of information on the Internet that started as fiction but has become 'fact' – just like 'Chinese whispers'!

- Sometimes you may look something up but the explanation appears too technical. Don't stop there; keep looking at alternative results produced by the search engine and you will invariably come across an explanation that is much simpler and easier to understand.

- Many sites have online tutorials that can help you understand how to work through a particular task, such as how to solder correctly or how to identify a dry joint. Go online now and see if you can find an interactive site.

Appendix

Appendix 1: Abbreviations

2D – Two-dimensional

3D –Three-dimensional

AC – Alternating current

A/C – Across corners

A/F – Across flats

Alum – Aluminium

ASSY – Assembly

BDMS – Bright drawn mild steel

BH – Brinell hardness number

BS – British Standard

BSI – British Standards Institution

CAD – Computer aided design

CBORE – Counterbore

CHAM – Chamfer

CI – Cast iron

CIM – Computer integrated manufacturing

CL – Centreline

CNC – Computer numerical control

CRMS – Cold rolled mild steel

CRS – Centres

CSK – Countersunk

Dural – Duralumin

DTI – Dial test indicator

EMF – Electromotive force

FSH – Full service history

GPR – Glass-reinforced plastic

HDMI – High definition multimedia interface

HEX HD – Hexagon head

I/D – Inside diameter

ISO – International Organisation for Standardisation

LED – Light emitting diode

LDR – Light dependent resistor

MDF – Medium density fibreboard

MB – Megabyte

Ø – Diameter (preceding a dimension)

O/D – Outside diameter

PAT – Portable appliance testing equipment

PCB – Printed circuit board

Phos Bronze – Phosphor-bronze

PPE – Personal protective equipment

PVC – Polyvinylchloride

PTFE – Polytetrofluoroethylene

PE – Potential energy

PCD – Pitch circle diameter

R – Radius (preceding a dimension, capital only)

SG Iron – Spherodial graphite cast iron

SS – Stainless steel

SMT – Surface-mount technology

SWG – Standard wire gauge

THK – Thick

TIG – Tungsten inert gas welding

TYP – Typical or typically

USB – Universal serial bus

VA – Volt-amperes

VPN – Vickers pyramid hardness number

Appendix 2: Standard symbols for electronic components

Connected wires

Unconnected wires

Cell

Battery or cells

Earth

Fuse

Transformer

DC supply

AC supply

Lamp

Resistor

Variable resistor

Variable capacitor

Capacitor with pre-set adjustment

Voltmeter

Ammeter

Switch

Capacitor

Capacitor (polarised)

LED

Diode

Amplifier

Bell

Buzzer

Transistor

Appendix 3: Standard hydraulic symbols

Basic symbols

Pump or motor

Measuring device

One square-pressure – control;

two or three adjacent squares – directional control

Conditioning apparatus such as a filter, heat exchanger, separator or lubricator.

Spring

Restriction (affected by viscosity)

Restriction unaffected by viscosity

Direction of hydraulic Fluid

Direction of pneumatic flow or exhaust to atmosphere.

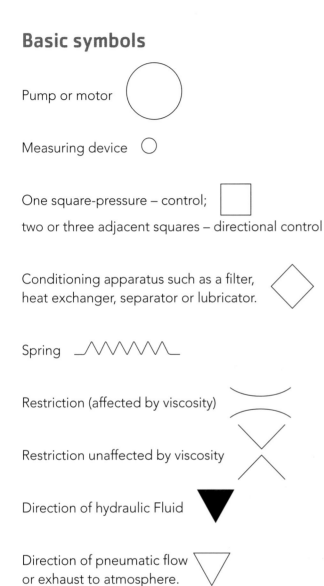

Energy conversion

Fixed capacity hydraulic pumps (convert hydraulic or pneumatic energy into rotary mechanical energy.)

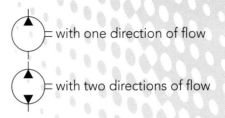

with one direction of flow

with two directions of flow

Motors (convert hydraulic or pneumatic energy into rotary mechanical energy.)

Fixed capacity hydraulic motor with one direction of flow

Oscilliating motor with two directions of flow

Directional control valves (provide full or restricted flow by opening or closing of one or more flow paths.)

Flow paths:

One flow path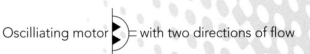

Two closed ports

Two flow paths

Two flow paths and one closed port

Two flow paths with cross connection

One flow path in a by-pass position, two closed ports

Reservoirs

Reservoir open to atmosphere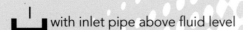

 with inlet pipe above fluid level

with inlet pipe below fluid level

with a header line

Pressurised reservoir

Filter or strainer

Non-return valve:

free (opens if the inlet pressure is higher than the outlet pressure)

spring loaded (opens if the inlet pressure is greater than the outlet pressure)

Sources of energy

Pressure source

Electric motor

Heat engine

Control methods

Muscular control:

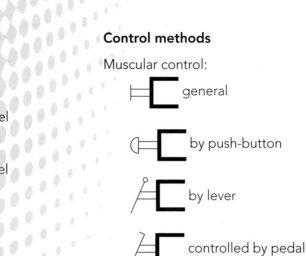 general

by push-button

by lever

controlled by pedal

Mechanical control:

 by plunger or tracer

 by spring

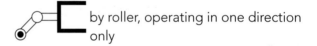 by roller

by roller, operating in one direction only

Appendix 4: Information on selected engineering materials

Table A.1: Ferrous metals

Material	Carbon content and other elements	Properties	Applications
Mild steel	0.1–0.3% carbon	Strong, fairly malleable and ductile	Wire, rivets, nuts and bolts, pressings, girders; general workshop material
Medium carbon steel	0.3–0.8% carbon	Strong and tough, can be hardened by heat treatment	Hammer heads, cold chisels, gears, couplings; impact-resistant components
High carbon steel	0.8–1.4% carbon	Strong, tough, and can be made very hard by heat treatment	Knives, springs, screw-cutting taps and dies; sharp-edged tools
Grey cast iron	3.2–3.5% carbon	Weak in tension but strong and tough in compression; very fluid when molten	Lathe beds, brake drums, engine cylinder blocks and cylinder heads, valve bodies
Stainless steel	Up to 1.0% carbon Up to 27% chromium Up to 0.8% manganese	Corrosion resistant, strong, tough	Food processing and kitchen equipment, surgical equipment, decorative items

Table A.2: Non-ferrous metals

Material	Composition	Properties	Applications
Copper	Almost pure	Very ductile and malleable; good conductor of heat and electricity; corrosion resistant	Electrical wire and cable, water pipes, soldering iron bits, alloying to make brasses and bronzes
Zinc	Almost pure	Soft, rather brittle, good fluidity when molten; corrosion resistant	Protective coating; alloying to make brasses
Tin	Almost pure	Very soft and malleable; highly corrosion resistant	Protective coating; alloying to make solders
Lead	Almost pure	Extremely soft, heavy and malleable; highly corrosion resistant	Roofs, lining of tanks; alloying to make solders
Aluminium	Almost pure	Soft, light and malleable; corrosion resistant	Wide range of domestic products and containers; wide range of alloys
Brasses	Up to 70% copper Up to 40% zinc Up to 1% tin	Very ductile with high copper content; very strong, tough, and fluid when molten with high zinc content	Tubes, pressings, forgings and castings for a wide range of engineering and marine components

Material	Composition	Properties	Applications
Tin-bronzes	Up to 96% copper Up to 22% tin Up to 2% zinc Up to 0.5% phosphorus	Very malleable and ductile with high copper content; very strong, tough and fluid when molten with high tin content	Springs, electrical contacts, bearings, gears, valve and pump components
Aluminium alloys	Up to 97% aluminium Up to 5% silicon Up to 3% copper Up to 0.8% magnesium Up to 0.8% manganese	Ductile, malleable with good strength, and good fluidity when molten	Electrical power lines, ladders, aircraft and motor vehicle components, light sand and die castings

Table A.3: Common thermoplastic materials

Common name	Properties	Applications and uses
Low-density polythene (LDPE)	Tough, flexible, solvent resistant: degrades if exposed to light or ultraviolet radiation	Flexible squeeze containers, packaging, piping, cable and wire insulation
High-density polythene (HDPE)	Harder and stiffer than low-density polythene, with higher tensile strength	Food containers, pipes, mouldings, tubs, crates, kitchen utensils, medical equipment
Polypropene (PP)	High tensile strength and a high melting point; can be produced as a fibre	Tubes, pipes, fibres, ropes, electronic components, kitchen utensils, items of medical equipment
PVC	Can be made tough and hard or soft and flexible; solvent resistant; soft form tends to harden with time	When hard, window frames, piping and guttering; when soft, cable and wire insulation, upholstery
Polystyrene	Tough, hard, rigid but somewhat brittle; can be produced as a light cellular foam; liable to be attacked by petrol-based solvents	Foam mouldings used for packaging and disposable drinks cups; solid mouldings. Used for refrigerator mouldings and other appliances in its solid form
Perspex	Strong, rigid, transparent, but easily scratched; easily softened and moulded; can be attacked by petrol-based solvents	Lenses, corrugated sheets for roof lights, protective shields on machinery, aircraft windows, light fittings
PTFE	Tough, flexible, heat-resistant, highly solvent resistant; has a waxy, low-friction surface	Bearings, seals, gaskets, non-stick coatings for kitchen utensils, tape
Nylon	Tough, flexible and very strong; good solvent resistance, but does absorb water, and deteriorates with outdoor exposure	Bearings, gears, cams, bristles for brushes, textiles
Terylene	Strong, flexible and solvent resistant; can be made as a fibre, tape or sheet	Reinforcement in rubber belts and tyres, textile fibres, recording tape, electrical insulation tape

Table A.4: Common thermosetting plastics

Common name	Properties	Applications and uses
Bakelite	Hard; resistant to heat and solvents; good electrical insulator and machinable, colours limited to brown and black.	Electrical fittings and components, vehicle distributor caps, heat resistant handles
Formica	Similar properties to bakelite but naturally transparent, and can be produced in a variety of colours	Electrical fittings, bathroom fittings, kitchenware, trays, laminates.
Melamine	As above, but harder, and with better resistance to heat; very smooth surface finish when moulded or machined	Electrical equipment, tableware, control knobs, handles, laminates
Epoxy resins	Strong, tough, good chemical and thermal stability; good electrical insulator, good adhesive	Glass and carbon fibre reinforced panels for vehicles, flooring material, laminates, adhesives
Polyester resins	Strong, tough; good wear resistance, and resistance to heat and water. Good electrical insulator	Boat hulls, motor panels, aircraft parts, fishing rods, skis, laminates

Appendix 5: SI unit prefixes and symbols

Table A.5

Multiple	Prefix	Symbol
10^{12}	tera	T
10^{9}	giga	G
10^{6}	mega	M
10^{3}	kilo	k
10^{2}	hecto	h
10^{1}	deka	da
10^{-1}	deci	d
10^{-2}	centi	c
10^{-3}	milli	m
10^{-6}	micro	μ
10^{-9}	nano	n
10^{-12}	pico	p
10^{-15}	femto	f
10^{-18}	atto	a

Appendix 6: SI derived units

Table A.6

Physical quanity	Name	Symbol
Electric charge	coulomb	C
Electric capacitance	farad	F
Electrical inductance	henry	H
Frequency	hertz	Hz
Energy, work, amount of heat	joule	J
Illuminance	lux	Lx
Force, weight	newton	N
Electric resistance	ohm	Ω
Pressure, stress	pascal	Pa
Electric conductance	siemens	S
Magnetic flux density	tesla	T
Electromotive force	volt	V
Power, radiant flux	watt	W
Induction magnetic flux	weber	Wb

Appendix 7: Greek alphabet

Alpha	A	α
Beta	B	β
Gamma	Γ	γ
Delta	Δ	δ
Epsilon	E	ε
Zeta	Z	ζ
Eta	H	η
Theta	Θ	θ
Iota	I	ι
Kappa	K	κ
Lambda	Λ	λ
Mu	M	μ
Nu	N	ν
Xi	Ξ	ξ
Omicron	O	ο
Pi	Π	π
Rho	P	ρ
Sigma	Σ	σ
Tau	T	τ
Upsilon	Υ	υ
Phi	Φ	φ
Chi	X	χ
Psi	Ψ	ψ
Omega	Ω	ω

Glossary

Absolute value – the numerical value of a number regardless of the sign before it.

Accident – an unforeseen event causing injury or damage.

Additional view – many engineering drawings have complex features that cannot be *seen* easily using conventional orthographic projection techniques. You can produce an extra view or partial view, sometimes shown at a larger scale, to allow clear details to be seen. CAD allows you to quickly copy and scale elements to produce these additional views.

Adjacent – the side next to the angle.

Alarm circuit – a circuit in which a single or combination of input conditions 'triggers' the circuit into switching states. Used in intruder alarm systems, smoke alarms, car alarms and even low-battery indicators. These circuits use transistors, combinational logic, or a mixture of both.

Alloy – a metal that consists of two or more metals mixed together.

Ampere – the SI unit of electric current.

Alternating current – often referred to as a.c.; electrical power that is not at a fixed level but alternates between a positive peak and a negative peak. The common waveform for alternating current is a sine wave. The mains supply is a.c.

Archiving – cataloguing and storing data that has accumulated over a period of time.

Armature – anything that moves under magnetic influence.

Assembly point – a place of safety; a signposted position outside a building, usually in a car park or on a sports field.

Base – when we raise a number to a power we call the number the base and the power the index.

Base point – When you draw or modify objects, the base point is the key point of reference. For example, if you were drawing a circle, the centre of the circle would probably be your base point. Similarly, when you rotate an object you need a pivot or base point around which to rotate it.

Blockboard – a type of laminate: It consists of strips of wood bonded together and sandwiched between two thin outer sheets.

Carcinogenic – a substance that can cause cancer if inhaled, ingested or if it penetrates the skin.

Case hardening – the process of hardening the surface of a metal.

CD-ROM – compact disc, read-only memory.

Chipboard – a cheap, hard material that consists of resin bonded with wood particles that originate from recycled waste such as saw dust, wood shavings and substandard pieces that have been reduced to particles and fibres.

CIM – computer integrated manufacture. The entire manufacturing process is controlled by computer. This allows real-time information to be passed easily between departments involved with design, planning, purchasing of raw materials, manufacturing, quality assurance and other business functions.

Coefficient of kinetic friction – the force needed to keep an object moving is slightly less than that needed to overcome static friction. This is called the kinetic frictional resistance and the slightly lower value of μ that results is called the coefficient of kinetic friction.

Coefficient of static friction – the ratio of the force of friction acting between two surfaces in contact. Its symbol is the Greek letter μ (mu).

Combinational logic circuit – logic circuit in which the output depends directly on the state of the inputs. Used for circuits where you know the output you require for known input conditions: for example, NOT, AND, NAND, OR, NOR and EXOR gates.

Commutator – a segmented rotating metal cylinder that contacts the brushes of a direct-current motor to provide a unidirectional current from the generator or a reversal of current into the coils of the motor.

Competent person – a trained employee who has the knowledge, skills, ability, experience and positive attitude for a specific task.

Computer network – a group of computers that are connected to a server, which is a central point for storing files and software. It allows controlled access to the network with passwords and login profiles for security, so that users can share files and folders.

Concurrent – existing or happening at the same time.

Constant of proportionality – two quantities are called proportional if they vary in such a way that one of the quantities is a constant multiple of the other, or equivalently if they have a constant ratio.

Contactors – a device for switching high-voltage power supply systems on or off.

Corrosive – a substance that destroys living tissues.

Cosine – the measurement of an acute angle in a triangle with a right angle, that is calculated by dividing the length of the side next to it by the length of the hypotenuse.

Coulomb (C) – a unit of electrical charge which measures the amount of electricity transported in one second by a current of one ampere.

Cube – of a number is the number multiplied by itself twice.

Datum – an edge, surface, centre, point, line or other feature that can be used as a reference to measure from.

Deburring – a finishing operation to remove burrs, jagged and raised edges or surface imperfections; usually performed with careful use of a file.

Decay – when the decrements in one variable produce smaller decrements in the other.

Dependent variable – a variable whose value depends on the independent variable.

Direct current – often referred to as d.c.; electrical power that is at a fixed level. Cells and batteries deliver direct current. Mains-powered battery chargers and adapters for small devices such as mobile phones and MP3 players contain a circuit to convert a.c. to d.c.

Down-cut – a method of removing material where cutter rotates in the same direction as the table feeding the work.

Drawing copper wire – copper is very ductile and can be progressively reduced in diameter, by drawing it through dies, to various sizes for wires and cables.

e-format – information displayed on a computer screen.

Electrical power – the rate of energy transfer. Its SI unit is the watt (W).

Electrical shock –the passage of electric current through part of the body.

Electromagnetic induction – the process by which voltage is produced in a coil as a result of a changing magnetic field in the circle formed by the loop.

Electromotive force (emf) – the force which causes motion of the electrical charges.

Emergency evacuation – process used during a fire, gas release or other emergency to ensure that everyone affected leaves the building to a place of safety, usually outside and designated as an assembly point. The evacuation is initiated by the sounding of an alarm such as a two-tone siren

Emergency lighting – the green and white signs that indicate the evacuation route in an emergency.

Employee – a person who is employed, through a contract of employment, to carry out specific jobs, tasks or activities for an employer. Examples: designers, maintenance fitters, electricians.

Employer – the person/s that run, control and manage a workplace. Examples: a factory, a manufacturing plant, a construction site.

Equilibrant E – the equilibrant is the additional force that must be applied equal and opposite to the resultant, to keep the body in a state of equilibrium.

Escape route – used during evacuation. The escape route is designated by green and white signs, showing a white arrow to indicate the direction, a walking person and sometimes a door. This sign category is a safe condition because it shows the route to a safe place.

Etching – The PCB starts life as a board with a coating of copper on one or both sides. The interconnecting wires are traced onto the copper using etch resist – something that will protect the copper and stop it being removed. The board is then placed in an etching solution that removes all the unprotected copper, leaving only the copper tracks. This process is called etching.

Explosive – able or likely to explode.

Exponential decay – when regular decrements in one variable produce ever smaller decrements in the other.

Exponential growth – when regular increments in one variable produce increasingly large increments in the other.

Fire warden – someone trained to ensure that everyone evacuates a building during an emergency evacuation. Their role is to ensure that the evacuation is carried out in an orderly manner and to check the building for any persons who have not responded.

First-aider – someone who has been trained in first aid and is listed as having that extra responsibility. Their role is to deal with emergencies, to try to preserve life and prevent situations worsening until medical help arrives. They also treat minor injuries such as cuts and abrasions.

Fixtures and fittings – such items as electrical wiring, lighting, doors, ventilation systems, windows, carpets.

Flammable – a substance that burns easily.

Flux density – this is the number of lines of force (Wb) passing at right angles through an area of 1 square metre.

Flux linkage – property of a coil of conducting wire and the magnetic field through which it passes.

Forging – hot-working process in which red-hot pieces of steel are placed in the cavity of a forging tool and then shaped – into a spanner, for example.

FSH – Full Service History

Fuse – a thin wire conductor designed to fail and thus protect a circuit when the current passing through it exceeds a prescribed value. The fuse is said to 'blow', because the higher current causes the fuse wire to melt, thus breaking the supply to the circuit.

Gauge pressure (pg) – the difference in pressure between a system and the surrounding atmosphere

GRP – glass-reinforced plastic.

Hard copy – information printed on paper.

Hardboard – made from wood fibres and glued under heat and pressure. This is stronger and harder than chipboard as the wood fibres are highly compressed.

Hardening – mild steel components are sometimes given a hard surface layer by this process.

Hardware control measure at the point of the hazard – safety measures on machinery to protect the user such as guarding, extraction hoods and electrical switching.

Hazard – something with the potential to cause harm to you or to someone else.

HDMI – high-definition multimedia interface: a compact audio/video interface for transmitting uncompressed digital data. Set to become the next big seller in connection technology.

Hypotenuse – the longest side of a triangle that has a right angle

Independent variable – a variable whose value is not dependent on any other in the equation.

Induction – the first few days of starting at a new place of work.

Irritant – a substance that causes inflammation on contact with skin.

Inversely proportional – means that, as one quantity increases, the other one decreases by a proportional amount.

Isometric circles – when constructing circles in isometric views, CAD systems have a special tool for constructing isometric ellipses, although this is often an ellipse option.

Lathe – a machine that grips a piece of material and rotates it while tools of various types are applied to remove material.

Machining centre – a multi-function machine on which complex milling, drilling and boring operations are performed.

Magnetic flux – imaginary lines of force.

Magnetomotive force – the current flowing through the coil in a magnetic circuit multiplied by the number of turns. This is analogous to electromotive force in an electrical circuit.

Maintenance – routine activities carried out to prevent or reduce machinery breaking down. Breakdowns cause production stoppages.

Maintenance personnel – mechanical, electrical and fluid power specialists.

Manual handling – the transporting or supporting of a load (including lifting, putting down, pushing, pulling, carrying, or moving) by hand or by bodily force.

MDF (Medium Density Fibreboard) – manufactured by a dry process at a lower temperature than hardboard and with a different bonding agent. It is used for many applications such as furniture, shelves and doors as it is very smooth and does not swell in high humidites.

Memory metals – see Shape memory alloys.

Mutual induction – The process by which an emf is induced in the second coil as a result of the changing flux linkage between the coils.

Natural growth – when the figures of exponential growth are plotted on a graph.

Near-miss or incident – an unforeseen event that narrowly misses causing injury or damage. Example: dropping a spanner off an overhead crane and just missing somebody.

Ohm – SI unit of electrical resistance.

Opposite – the angles that are directly opposite each other where four lines meet.

Orientation – the position of an object in relation to a given reference or datum point. For example, a DVD has to be placed the correct way up in a computer drive.

Origin – the point where the x axis and y axis cross.

Orthographic projection – the representation of a three-dimensional object drawn on a two-dimensional surface using a series of linked views.

Oxidising – the combination of a substance with oxygen. Substances such as these may increase the risk and violence of a fire significantly if they come into contact with a flammable or combustible substance.

Permit to work – document issued to control entry into confined spaces. Issued by a senior engineer, for one activity at a time, to a competent engineer who is to carry out the work. The document indicates details of the work to be done and the safety precautions and is timed, usually for eight hours.

Pictorial representation – a technical illustration that shows the faces of a three-dimensional object in a single view drawn on a two-dimensional surface.

Piezoelectric effect – the application of a force to certain crystals that causes a potential difference to be set up across faces at right angles to the force.

Plain carbon steels – a steel in which iron and carbon are the main constituents.

Plastic injection moulding – process in which plastic pellets are fed into the cavity in a hot moulding tool to form products such as plastic spoons.

Plywood – the main type of laminated board. It consists of thin layers of wood bonded together, with their grain directions running alternately at right angles.

Polarity – the direction of electron flow from a power source such as a power supply, a cell or a battery. Electrons flow from the negative terminal through the circuit to the positive terminal. It is therefore important to connect a circuit using the correct polarity.

Polymer materials – long intertwined chains of molecules consisting of hydrogen and carbon atoms with others attached, which gives them their different properties. Such materials include plastic and rubber.

Potential difference – the change in voltage due to energy being used to pass current through some resistance.

PPE – personal protective equipment, worn as appropriate for the task being carried out. Examples: overalls, protective footwear (such as steel toe caps), eye protection, dust masks and respirators to cover the mouth and nose

Pre-use inspection – term used for the inspection and checking of tools or work equipment before use. It is good practice before starting an engineering activity.

Primary coil – one of two coils in a transformer which form the iron core. The primary coil is connected to the a.c. supply or input voltage V.

Principle of moments – the term used to describe the conditions for static equilibrium where the sum of moments of the forces acting on a body are zero, so that there is no rotation clockwise or anticlockwise.

Properties – the qualities or power that a substance has.

Proportion – the different amounts that a quantity is divided into.

Proportional relationship – a relationship between two variables which is represented on a graph by a straight line passing through the origin.

Prototype – the pre-production version of an electronic circuit or circuit board. The prototype is used for fully testing the final design or construction before the circuit is committed to mass production using a printed circuit board. You do not want to discover a problem with your circuit after the PCBs have been produced, so the prototype stage is an extremely important one if you want to avoid costly mistakes.

Quality control – methods and procedures for measuring, recording and maintaining quality targets.

Ratio – compares the sizes of the amounts that a quantity is divided into.

Reciprocal – the reciprocal of a number is the number divided into 1. That is to say it is made into a fraction with 1 on the top line and the number underneath it.

Rectangular hyperbola – the curve that results when inversely proportional quantities are plotted on a graph.

Relative density – also called specific gravity. This is the ratio of the density of one substance in relation to the density of another. This term is often used in reference to the density of a substance in comparison with water.

Relay – is a magnetically operated switch. It enables a small electric current in one circuit to switch a much larger electric current on or off in another circuit. A relay consists of a coil of wire wound on a soft iron core, a yoke and a hinged armature

Reluctance – analogous to resistance in an electrical circuit. but relating to a magnetic field and the resulting magnetic flux. The quantity of reluctance depends on the material on which the coil is wound.

Resolution – the motion of a force into two or more which have different directions.

Resultant – the resultant of a system of forces is their combined pulling effect on a body.

Risk – the likelihood that harm will occur and the severity of the harm from contact with the hazard.

Risk assessment – a five-step approach to evaluating hazards. Risk assessments should be carried out by health and safety representatives or trained persons. There are five steps to a risk assessment:

1. Look for the hazards.
2. Decide who is likely to be harmed, and how.
3. Evaluate the risks, and decide whether sufficient precautions have been taken or whether further action is required.
4. Record your findings.
5. Review your assessment, and make revisions if necessary.

Risk control systems – information, instruction, training, supervision to all employees and others affected by the engineering activity.

5S – a method for organising a workplace. It is based on a Japanese principle. The five S-words can be translated as follows:

Seiri (sort) – Sort out all tools and equipment and discard those that you don't need.

Seiton (set in order) – Arrange tools and equipment close at hand, using shadow boards, for example.

Seiso (sweep) – Keep the workplace and tools and equipment clean and neat. At the end of the session make sure everything is returned to where it belongs.

Seiketsu (standardise) – Ensure activities are carried out in a standard, repeatable way. This is important in manufacturing operations.

Shitsuke (sustain) – Perhaps the most important element is to keep things neat and tidy once the work area has been sorted out and not let bad habits build up.

Safe system of work – a step-by-step procedure to ensure the safe mix of the employee, machinery and materials used, level of training, instruction and supervision.

Scientific notation – a convenient way of representing numbers that maybe very large or very small. An alternative term for standard form.

Secondary coil – one of two coils in a transformer which form the iron core. The secondary coil delivers the required output voltage V_o.

Sensitising – adverse effects caused by a reaction to inhalation or penetration of a substance.

Servicing – routine work on machinery. Examples: oiling and greasing moving parts to prevent wear; changing filters that are partially blocked.

Shape memory alloy – also called memory metals. When deformed they will return to their original shape when heated, or when the forces are removed.

Sine – the fraction calculated for an angle by dividing the length of the side opposite it in a triangle that has a right angle, by the length of the the side opposite the right angle.

Smart materials – a material that can undergo a property change when there is a change in its working environment.

Solder – a soft metal alloy of lead and tin. It has a low melting point and is used for joining electrical wires and components together. The solder contains flux, which helps to clean the surfaces to be joined and protects the joint once the solder has set.

Solenoid – a current-carrying coil produces a magnetic field similar to that of a bar magnet. The coil is known as a solenoid.

Specialist input devices – computer peripherals are devices attached to the computer such as a printer, digitiser, scanner, or digital camera.

Square – of a number is the number multiplied by itself.

Square law – one quantity is proportional to the square of the other.

Standard form – also called scientific notation.

Static equilibrium – when a body (such as a cone) is at completely at rest we say that it is in a state of static equilibrium.

Surface finish – the surface finish of an object tells us how smooth its surface is. If you run the tip of your finger nail over a surface you will get an indication of how smooth it is. We can visualise the measure we use for surface finish as being like the distance between the high points and low points of a saw blade. Surface finish is measured in micrometres (μm): 1 μm = 0.000001 m.

Surface texture – the roughness or waviness of a surface.

Tangent – a straight line that touches the outside of a curve but does not cut across it.

Template – a standard layout that CAD operators use for producing drawings. It includes a border, a title block, and a standard logo for the company or organisation. It also features standard settings for such things as units, types of line and colours.

Tesla – a unit of magnetic flux density.

Thermoplastics – a material that softens when heated and hardens when cooled.

Thermosetting – the property that describes a material becoming permanently rigid when heated.

TIG – tungsten inert gas welding. The weld area is shielded by an inert gas, such as argon, to prevent atmospheric contamination of the hot surfaces.

Tinning – applying a coating of solder to keep the tip of the soldering iron clean and free from oxide and to aid heat transfer.

Tolerance – an allowable deviation from the desired size (no size can be achieved exactly).

Tool change – When machining on a lathe you will often need to perform a series of operations. You will need different tools for removing large amounts of material (roughing), for producing a precise surface finish (finishing), and for specialised operations such as facing and thread-cutting. This means that, each time, you will need to stop the machine, remove the guards, remove the tool, insert a different tool, secure the tool and replace the guards.

Torque – the output as a result of the force multiplied by the turning radius.

Toxic – a substance containing poison or caused by poisonous substances.

Toxic substances – chemicals or materials that are poisonous and need to be handled correctly. Some may give off fumes, so should be used in a well-ventilated area; others could be corrosive to the skin.

Traceability – the ability to trace parts and products back to their source.

Transistor – a device for controlling and amplifying current. There are two types (PNP and NPN). Both have three connections: base, emitter and collector. The field effect transistor (FET) also has three connections: gate, drain and source. All are used in amplifiers, audio, switching, optical and oscillator circuits.

Transposition – depending on which of the physical quantities we need to solve a problem, we may need to change a formula round to make that quantity the subject. The process is called transposition.

Up-cut – a method of material removal where the cutter rotates in the opposite direction to the table feeding the work.

USB – universal serial bus: a standard specification for connecting computers and peripheral devices.

Vector addition – a measurement based on combining the magnitude and direction of individual vectors.

Vernier protractor – used for measuring angles to a high degree of accuracy. More accurate than the combination square and can incorporate a scribing point.

Volt – SI unit of electromotive force.

Weber – the SI unit of magnetic flux.

Work equipment – covers any type of equipment, from a simple hand tool to the most complex piece of machinery. Examples: from spanners and hammers to the complex assembly equipment used in car manufacture.

Workpiece – any component or piece of material that is to be machined or worked upon to create a final product.

Zener – a diode that makes specific use of what is called the reverse bias effect. It can be designed to begin conducting at a specific reverse bias voltage and so is used for stabilising or regulating a voltage source.

Index